Lecture Notes in Mathematics 1782

Editors:
J.-M. Morel, Cachan
F. Takens, Groningen
B. Teissier, Paris

Springer
Berlin
Heidelberg
New York
Barcelona
Hong Kong
London
Milan
Paris
Tokyo

Cho-Ho Chu   Anthony To-Ming Lau

Harmonic Functions on Groups and Fourier Algebras

Springer

Authors

Cho-Ho Chu
Goldsmiths College
University of London
London SE14 6NW
United Kingdom
e-mail: c.chu@gold.ac.uk

from September 2002:
Queen Mary College
University of London
London E1 4NS
United Kingdom

Anthony To-Ming Lau
Department of Mathematical
and Statistical Sciences
University of Alberta
Edmonton, Alberta
T6G 2G1 Canada
e-mail: tlau@math.ualberta.ca

Cataloging-in-Publication Data applied for
Die Deutsche Bibliothek - CIP-Einheitsaufnahme

Chu, Cho-Ho:
Harmonic functions on groups and Fourier algebras / Cho-Ho-Chu ; Anthony
To-Ming Lau. - Berlin ; Heidelberg ; New York ; Barcelona ; Hong Kong ;
London ; Milan ; Paris ; Tokyo : Springer, 2002
 (Lecture notes in mathematics ; 1782)
 ISBN 3-540-43595-6

Mathematics Subject Classification (2000):
43A05, 43A20, 43A35, 31C05, 45E10, 22D25, 46L70, 32M15

ISSN 0075-8434
ISBN 3-540-43595-6 Springer-Verlag Berlin Heidelberg New York

Springer-Verlag Berlin Heidelberg New York a member of BertelsmannSpringer
Science + Business Media GmbH

http://www.springer.de

© Springer-Verlag Berlin Heidelberg 2002
Printed in Germany

Typesetting: Camera-ready TEX output by the author

SPIN: 10866636   41/3142/du - 543210 - Printed on acid-free paper

Preface

The purpose of this monograph is to introduce some new aspects to the theory of harmonic functions and related topics. They are a fusion of some recent developments in non-associative functional analysis, semi-groups and harmonic analysis. More specifically, we study the algebraic analytic structures of the space of bounded *complex* harmonic functions on a locally compact group G and its non-commutative analogue, the space of harmonic functionals on the Fourier algebra $A(G)$. We show that they are both the ranges of contractive projections on von Neumann algebras and therefore admit Jordan algebraic structures which are usually non-associative. This provides a natural setting to apply new methods and results from non-associative analysis, semigroups and Fourier algebras. We use these devices to study, among others, the Poisson representation of bounded complex harmonic functions on G, the semigroup structures of the Poisson space and the non-associative geometric structures of the harmonic functionals.

This work was done during several mutual visits of the authors at the University of Alberta and University of London, supported by EPSRC and NSERC grants. All results are new, some of which have been presented at seminars and workshops in London, Oxford, Edmonton, Hong Kong, Irvine, Toulouse and Oberwolfach. We thank warmly the audience at these institutions for their inspiration and hospitality. Above all, we are grateful to our families for their constant support and encouragement.

Key words and phrases: Locally compact group. Harmonic function. Liouville property. Poisson representation. Compact semigroup. Almost periodic function. Distal function. Harmonic functional. Fourier algebra. Group von Neumann algebra. Banach algebra. Arens product. C*-algebra. Jordan algebra. JB*-triple.

Work supported by EPSRC grant GR/M14272 and NSERC grant A7679

Contents

1. Introduction

Let G be a Lie group and let Δ be the Laplace operator on G. A function $f \in C^\infty(G)$ is *harmonic* if $\Delta f = 0$. It is well-known in this case that there exists a family $\{\sigma_t\}_{t>0}$ of absolutely continuous probability measures on G such that f satisfies the following convolution equations

$$f = \sigma_t * f \qquad (t > 0)$$

which motivates the following definition.

Let σ be a probability measure on a locally compact group G. A real Borel function f on G satisfying the convolution equation

$$f(x) = \sigma * f(x) = \int_G f(y^{-1}x)d\sigma(y) \qquad (x \in G)$$

is called *σ-harmonic*. Harmonic functions on groups have been widely studied for a long time and play important roles in analysis, geometry and probability theory. Naturally one can also consider *complex* measures σ and the *complex* σ-harmonic functions on G defined as above. The complex case, however, does not seem to have been studied in as much detail and depth as the real case.

In this research monograph, we give a functional analytic approach to the *complex* case by studying the algebraic and analytic structures of the space of bounded complex σ-harmonic functions on a locally compact group G and its non-commutative analogue, that is, the space of harmonic functionals on the Fourier algebra $A(G)$. We show that both spaces are the ranges of contractive projections on von Neumann algebras and therefore admit Jordan algebraic structures. This introduces a new aspect, namely, the non-associative algebraic and geometric structures, to the theory of harmonic functions and harmonic functionals. We exploit these structures to study the Poisson representation of bounded complex harmonic functions, the semigroup structures of the Poisson space and the non-associative geometric structures of the harmonic functionals.

The first part of the monograph is concerned with complex harmonic functions and is contained in Chapter 2. Harmonic functionals are discussed in Chapter 3 which is the second part of the monograph.

To outline our main ideas, we first review briefly some relevant background of real harmonic functions, more details will be given later. For a probability measure σ on a locally compact abelian group G, it is a well-known result of Choquet and Deny [11,21] that every bounded continuous σ-harmonic function f on G is periodic and every point in the support of σ is a period, in particular, if the support of σ generates a dense subgroup of G, then f must be constant. Choquet and Deny have also shown that the positive (unbounded) σ-harmonic functions on an abelian group G are integrals of exponential functions, given that the support of σ generates a dense subgroup. Similar results also hold for nilpotent groups [4,14,22,36,45]. Recently, results of this kind have been extended to other classes of groups (see, for example, [17,18,66]). On the other hand, given an absolutely continuous probability measure σ on a semisimple Lie group G with finite centre, Furstenberg [27] has shown that the space of bounded uniformly continuous σ-harmonic functions on G forms an abelian C^*-algebra in certain product from which he derived a Poisson representation of bounded harmonic functions and investigated the deep issue concerning the boundary theory of G. In fact, Furstenberg's construction of the abelian C^*-product is valid for probability measures on arbitrary locally compact groups, as shown by Azencott [5], and this is the starting point of our investigation for the complex case.

Let σ be a complex measure on a locally compact group G with $\|\sigma\| = 1$. As usual, the Lebesgue spaces of the Haar measure on G are denoted by $L^p(G)$ for $1 \leq p \leq \infty$. We study the space of bounded complex σ-harmonic functions on G via the duality of $L^1(G)$ and $L^\infty(G)$ in the following way. Let J_σ be the norm-closure of $\{\check{\sigma} * f - f : f \in L^1(G)\}$. Then J_σ is a closed right ideal of $L^1(G)$ and its annihilator $J_\sigma^\perp = \left(L^1(G)/J_\sigma\right)^*$ is the space $\{f \in L^\infty(G) : \sigma * f = f\}$ of bounded σ-harmonic functions on G. One can define a contractive projection $P : L^\infty(G) \longrightarrow J_\sigma^\perp$ which, by results of [26,59], induces an abelian C^*-algebraic structure on $J_\sigma^\perp$ and this structure differs from that of $L^\infty(G)$ in general. From this, we can derive, for absolutely continuous σ, a Poisson representation of $J_\sigma^\perp$ as well as its subspace $J_\sigma^\perp \cap C_{\ell u}(G)$ of (left) uniformly continuous harmonic functions, and study the semigroup structures of the Poisson space which is the spectrum of $J_\sigma^\perp \cap C_{\ell u}(G)$. The C^*-product we obtain coincides with the one defined in [5,27] mentioned above when σ is a probability measure. Therefore our construction gives a unified approach to the Poisson representation as well as some functional analytic insights. We will not, however, be concerned with the boundary theory for G, but focus our attention on the semigroup properties of the Poisson space which are obtained by a fundamental structure theorem on compact semigroups in [71] (see also [40, p.16] and [39, p.143]) and the details are given in Sections 2.3, 2.4 and 2.5. Further properties relating to almost periodic and distal functions are discussed in the latter two sections. Many of our results are new even when σ is a probability measure in which case Poisson spaces corresponding to non-degenerate absolutely continuous measures on a second countable group G are shown, in Theorem 2.6.1, to have the same semigroup structure if G acts transitively on them.

We should point out that, for a probability measure σ, Paterson [64] has also defined a contractive projection $P : C_{\ell u}(G) \longrightarrow J_\sigma^\perp \cap C_{\ell u}(G)$ which induces a C^*-product on $J_\sigma^\perp \cap C_{\ell u}(G)$. The product in this case is simpler because the constant-1 function is harmonic. For complex measures σ, $J_\sigma^\perp$ need not contain constant functions but the C^*-product can be defined in terms of an extremal function in $J_\sigma^\perp$ and its explicit form for $J_\sigma^\perp \cap C_{\ell u}(G)$ is given in Theorem 2.2.17. Paterson [64] has also studied the case of a compact topological semigroup G and the corresponding Poisson space for a non-degenerate probability measure σ. We note that Willis [83] has studied the Banach algebraic structures of $L^1(G)/J_\sigma$ in detail, for a probability measure σ.

There is a non-commutative analogue of harmonic functions in the context of Fourier algebras and group von Neumann algebras. The non-commutative version of the duality between $L^1(G)$ and $L^\infty(G)$ is that between the Fourier algebra $A(G)$ and the group von Neumann algebra $VN(G)$ of a locally compact group G. Given a complex function σ in the Fourier-Stieltjes algebra $B(G) \supset A(G)$ which acts on $VN(G)$, we define I_σ to be the norm-closure of $\{\sigma\phi - \phi : \phi \in A(G)\}$ in $A(G)$. Then I_σ is a closed ideal in $A(G)$ and its annihilator $I_\sigma^\perp = (A(G)/I_\sigma)^*$ is the space $\{T \in VN(G) : \sigma.T = T\}$ which we call the σ-harmonic functionals on $A(G)$. The reason for this terminology is that, if G is abelian and if $\sigma = \hat{\mu}$ is the Fourier transform of a measure μ on the dual group $\hat{G}$, then $I_\sigma^\perp$ identifies with $J_{\check{\mu}}^\perp \subset L^\infty(\hat{G})$ where $d\check{\mu}(x) = d\mu(x^{-1})$. For arbitrary groups G, we can therefore view $I_\sigma^\perp \subset VN(G)$ as a non-commutative analogue of $J_\sigma^\perp \subset L^\infty(G)$. The space $I_\sigma^\perp$ was first studied by Granirer in [31]. We show that, for $\sigma \in B(G)$ with $\|\sigma\| = 1$, there is a contractive projection $P : VN(G) \longrightarrow I_\sigma^\perp$ (see also [31]). In contrast to the case of $J_\sigma^\perp$, the space $I_\sigma^\perp$ need not form a C^*-algebra, but by results of [25, 48], P induces a Jordan algebraic structure on $I_\sigma^\perp$ which is therefore related to infinite dimensional holomorphy and gives many interesting consequences. The Banach algebraic properties of $A(G)/I_\sigma$ are derived in Section 3.2. Although $L^1(G)/J_\sigma$ is always an L^1-space, $A(G)/I_\sigma$ need not be a Fourier algebra. We give necessary and sufficient conditions for $A(G)/I_\sigma$ to be isometrically isomorphic to the Fourier algebra of a locally compact group. The Jordan and geometric structures of $I_\sigma^\perp$ are studied in Section 3.3. We show that $I_\sigma^\perp$ is a

JW^*-algebra and determine when it is a von Neumann subalgebra of $VN(G)$, and when it is a Jordan subtriple of $VN(G)$. We give in Theorem 3.3.12 an explicit form of the Jordan triple product in $I_\sigma^\perp$ which is the non-associative counterpart of the C^*-product in $J_\sigma^\perp$. We also explain briefly in this section the relationship between Jordan triple systems and symmetric manifolds. By [16], a connected component $\mathcal{M}$ of projections in $I_\sigma^\perp$ is a symmetric real analytic Banach manifold. As in [16], we define an affine connection on $\mathcal{M}$ and describe its geodesics. In Section 3.4, we study the Murray-von Neumann classification of $I_\sigma^\perp$ and we classify $I_\sigma^\perp$ by the geometric properties of its preduals. We show that the types of $I_\sigma^\perp$ are invariant under linear isometries.

It will be of significant interest to study other properties of the Banach algebras $L^1(G)/J_\sigma$ and $A(G)/I_\sigma$, such as cohomology and amenability, when J_σ is a two-sided ideal of $L^1(G)$. It will also be of interest to study further semigroup properties and harmonic analysis of the Poisson space of a complex measure on a locally compact group.

Recently, matrix-valued harmonic functions on groups have been studied in [15] and shown to form a ternary Jordan algebra, that is, a Jordan triple system, by the fact that they are the range of a contractive projection on a non-abelian von Neumann algebra. This provides a new setting to extend the results in this monograph as well as other well-known results on scalar-valued harmonic functions.

Finally, we hope that the connection between Jordan algebras and harmonic analysis can be developed further.

2. Harmonic functions on locally compact groups

In this Chapter, we study the space of bounded (left uniformly continuous) σ-harmonic functions on a locally compact group G with respect to a complex measure σ on G. We first show that, for $\|\sigma\| = 1$, it is the range of a contractive projection on $L^\infty(G)$ and hence it has an abelian von Neumann algebraic structure. Further we show that the von Neumann algebra product is given by

$$(f \times g)(x) = \lim_\alpha \int_G f(y^{-1}x)\overline{u}(y^{-1}x)g(y^{-1}x)d\mu_\alpha(y) \qquad (x \in G)$$

where μ_α belongs to the convex hull of $\{\sigma^n : n \geq 1\}$ and u is an extreme point in the closed unit ball of the space. We derive a Poisson integral representation of the σ-harmonic functions on a Poisson space Π_σ and we show that Π_σ has a natural semigroup structure. Moreover, it is a compact left topological semigroup. Consequently its minimal ideal is isomorphic to a simple semigroup. We also discuss almost periodic and distal harmonic functions and related compactifications. We end the Chapter with some examples as well as supplementing Azencott's result [5] by showing that Poisson spaces corresponding to non-degenerate absolutely continuous measures on a second countable group have the same semigroup structure under transitive action of the group.

2.1. Preliminaries and notation

We denote throughout by G a locally compact group with identity e and the left invariant Haar measure λ. Unless otherwise stated, all groups are locally compact. By a *measure* on G, we mean a real or complex-valued regular Borel measure (of totally bounded variation).

Let $C_0(G)$ be the Banach space of complex continuous functions on G vanishing at infinity. Then its dual $C_0(G)^*$ identifies with all the complex regular Borel measures on G, also denoted by $M(G)$. For $\sigma \in M(G)$, its norm $\|\sigma\|$ is the total variation $|\sigma|(G)$. We denote by $M_+(G)$ the cone of positive measures in $M(G)$. Let $M^1(G) = \{\sigma \in M(G) : \|\sigma\| = 1\}$ and $M_+^1(G) = M^1(G) \cap M_+(G)$. Given $\mu \in M_+(G)$, its support is denoted by supp μ. We note that if μ is absolutely continuous (with respect to λ), then the group generated by supp μ is open.

A measure $\mu \in M(G)$ is called *non-degenerate* if the semigroup generated by supp $|\mu|$ is dense in G. If μ is non-degenerate, then it is *adapted* which means that supp $|\mu|$ generates a dense subgroup of G. The following lemma is known [41, p. 32], we include a proof for completeness.

Lemma 2.1.1. *Let G be a locally compact group with $\mu \in M(G)$. Then* supp$|\mu|$ *is σ-compact. If μ is adapted, then G is σ-compact.*

Proof. Write $S = $ supp$|\mu|$. By regularity of μ, there exist compact sets

$$S_1 \subset S_2 \subset \cdots$$

such that $S = \overline{\bigcup_{n=1}^{\infty} S_n}$. We can find open sets $U_1 \subset U_2 \subset \cdots$ such that $S_n \subset U_n$ with $\overline{U}_n$ compact. Let

$$G_n = \bigcup_{m=1}^{\infty} (U_n \cup U_n^{-1})^m = \bigcup_{m=1}^{\infty} (\overline{U}_n \cup (\overline{U}_n)^{-1})^m$$

where we note that $\overline{U} \subset U^2 U^{-1}$ for an open set U. Then (G_n) is an increasing sequence of σ-compact clopen groups with $G_n \supset S_n$. It follows from $S = \overline{\bigcup_{n=1}^{\infty} S_n} \subset \bigcup_{n=1}^{\infty} G_n$ that S is σ-compact. Finally, if μ is adapted, the above argument shows $G = \bigcup_{n=1}^{\infty} G_n$ which is σ-compact. $\square$

Lemma 2.1.2. *If an abelian group G admits an adapted measure $\mu \in M(G)$, then its dual group $\widehat{G}$ is first countable.*

Proof. The sets of the form

$$P(F, \varepsilon) = \{\chi \in \widehat{G} : |\chi(x) - 1| < \varepsilon \ \forall x \in F\}$$

where F is a compact subset of G and $\varepsilon > 0$, form a neighbourhood base of the identity $\iota \in \widehat{G}$ (cf. [38,p.361]). Let $G = \bigcup_{n=1}^{\infty} \bigcup_{m=1}^{\infty} \left(\overline{U}_n \cup (\overline{U}_n)^{-1}\right)^m$ be as in Lemma 2.1.1, where U_n is open and $\overline{U}_n$ compact. Then the finite intersections of the family

$$\left\{ P\left((\overline{U}_n \cup (\overline{U}_n)^{-1})^m, \frac{1}{k}\right) : m, n, k \in \mathbb{N} \right\}$$

form a countable neighbourhood base at $\iota \in \widehat{G}$. □

Given Borel functions $f, h : G \to \mathbb{C}$ and $\sigma \in M(G)$, we will adopt the following notation whenever it is well-defined:

$$(f * h)(x) = \int_G f(y)h(y^{-1}x)d\lambda(y)$$

$$(\sigma * f)(x) = \int_G f(y^{-1}x)d\sigma(y)$$

$$.(f * \sigma)(x) = \int_G f(xy^{-1})\Delta(y^{-1})d\sigma(y)$$

where Δ is the modular function on G. We also define

$$d\widetilde{\sigma}(y) = \Delta(y)d\sigma(y^{-1})$$

$$d\check{\sigma}(y) = d\sigma(y^{-1}) = \overline{d\sigma^*(y)} \, .$$

For the duality $\langle \cdot, \cdot \rangle : L^1(G) \times L^\infty(G) \to \mathbb{C}$, we have

$$\langle f * \sigma, h \rangle = \langle f, h * \widetilde{\sigma} \rangle$$

$$\langle \sigma * f, h \rangle = \langle f, \check{\sigma} * h \rangle$$

where $f \in L^1(G)$, $h \in L^\infty(G)$ and $\sigma \in M(G)$.

Let J_σ be the norm-closure of $\{\check{\sigma} * f - f : f \in L^1(G)\}$. Then we have $J_\sigma^\perp = \{h \in L^\infty(G) : \sigma * h = h\}$. We note that

$$\{f * \sigma - f : f \in L^1(G)\}^\perp = \{h \in L^\infty(G) : h * \tilde{\sigma} = h\}.$$

A Borel function $h : G \to \mathbb{C}$ is called σ-harmonic (or harmonic if σ is understood) if it satisfies the convolution equation

$$\sigma * h = h.$$

The space $J_\sigma^\perp$ of (essentially) bounded harmonic functions will be our main object of study in this chapter.

We first discuss briefly the case when $J_\sigma^\perp$ is trivial for a probability measure σ on G. The celebrated Choquet-Deny Theorem [11] states that, if G is abelian, then the bounded σ-harmonic functions are constant if (and only if) σ is adapted. For convenience, we say that G has the *Liouville property* if given any non-degenerate absolutely continuous probability measure σ on G, the bounded σ-harmonic functions on G are constant. It is well-known that the compact groups [51] and nilpotent groups [4,14,22,36] have the Liouville property. It has been shown in [45] that nilpotent groups of class 2 have the stronger Liouville property in that one need not assume absolute continuity of σ. It has also been shown in [17,18] recently that almost connected [IN]-groups have the Liouville property. In fact, it is shown in [18] that if σ is nonsingular with its translates, but not necessarily absolutely continuous, then every bounded σ-harmonic function on an [IN]-group G is constant on each connected component of G. We recall that a locally compact group G is called an [IN]-group if the identity has a compact invariant neighbourhood. A group G is called a [SIN]-group if every neighbourhood of the identity contains a compact invariant neighbourhood which is equivalent to the fact that the left and right uniform structures of G coincide. We note the following relevant result [46,70] which shows that $J_\sigma^\perp$ is never trivial if G is non-amenable and σ a probability measure. We recall that a locally compact group G is *amenable* if there is a positive norm one linear functional on $L^\infty(G)$ which is invariant under left translation. Solvable groups and compact groups are amenable, but the free group on two generators is not amenable.

Proposition 2.1.3. *Given a locally compact group G, the following conditions are equivalent:*

(i) *There is a probability measure μ on G such that the bounded μ-harmonic functions are constant;*

(ii) G *is amenable and* σ *-compact.*

Let $\widehat{G}$ be the dual space of G consisting of (equivalence classes of) nonzero continuous irreducible representations $\pi : G \to U(H_\pi)$ where $U(H_\pi)$ is the group of unitary operators on the Hilbert space H_π. For $\sigma \in M(G)$ and $f \in L^1(G)$, we define their Fourier transforms to be the following bounded operators on H_π :

$$\widehat{\sigma}(\pi) = \int_G \pi(x^{-1}) d\sigma(x); \quad \widehat{f}(\pi) = \int_G f(x)\pi(x^{-1}) d\lambda(x).$$

Lemma 2.1.4. *Let* σ *be a positive adapted measure on* G *and let* $\alpha = \sigma(G)$. *Then* α *is not an eigenvalue of* $\widehat{\sigma}(\pi)$ *for all* $\pi \in \widehat{G}\backslash\{\iota\}$ *where* ι *is the one-dimensional representation* $\iota(x) \equiv \text{Id}$.

Proof. We may assume $\alpha = 1$. Suppose otherwise, that $\widehat{\sigma}(\pi)\xi = \xi$ for some unit vector $\xi \in H_\pi$. Then

$$1 = \langle \widehat{\sigma}(\pi)\xi, \xi \rangle = \int_G \langle \pi(x^{-1})\xi, \xi \rangle \, d\sigma(x)$$

$$= \int_{\text{supp}\,\sigma} \text{Re} \, \langle \pi(x^{-1})\xi, \xi \rangle \, d\sigma(x).$$

If $\text{Re} \, \langle \pi(x^{-1})\xi, \xi \rangle < 1$ for some $x \in \text{supp}\,\sigma$, then the inequality holds in some open neighbourhood V of x by continuity and hence

$$1 = \int_V \text{Re} \, \langle \pi(x^{-1})\xi, \xi \rangle \, d\sigma(x) + \int_{G\backslash V} \text{Re} \, \langle \pi(x^{-1})\xi, \xi \rangle \, d\sigma(x)$$

$$< \sigma(V) + \sigma(G\backslash V) = 1$$

which is impossible. So we have $\langle \pi(x^{-1})\xi, \xi \rangle = 1$ for $x \in \text{supp}\,\sigma$ which gives $\pi(x^{-1})\xi = \xi$. It follows from the adaptedness of σ that $\pi(x)\xi = \xi$ for all $x \in G$ contradicting $\pi \neq \iota$. $\square$

We recall that a locally compact group G is called *central* if the quotient group G/Z is compact where Z is the centre of G. Central groups are unimodular and their irreducible representations are finite-dimensional. Given a central group G, it has been shown in [35] that $\widehat{G}$ admits a Plancherel measure η with the following Fourier inversion formula for $f \in L^1(G)$:

$$f(x) = \int_{\widehat{G}} (\dim \pi) \, \text{tr} \, (\widehat{f}(\pi)\pi(x)^*) d\eta(\pi) \tag{2.1}$$

whenever the function $\pi \mapsto (\dim \pi)\mathrm{tr}(\widehat{f}(\pi)\pi(x)^*) \in L^1(\widehat{G})$ for $x \in G$.

The following lemma is straightforward.

Lemma 2.1.5. *Given any locally compact group G with $\sigma \in M(G)$ and $\alpha \in \mathbb{C}$, the following conditions are equivalent for $f \in L^1(G)$:*

(i) $\sigma * f = \alpha f$;
(ii) $\widehat{f}(\pi)\widehat{\sigma}(\pi) = \alpha\widehat{f}(\pi)$ *for all $\pi \in \widehat{G}$.*

We give below a proof, which is different from [51] but generalizes to central groups, of the Liouville property of compact groups.

Lemma 2.1.6. *Let σ be an adapted probability measure on a compact group G and let f be a continuous function on G satisfying $\sigma * f = f$. Then f is constant.*

Proof. Since G is compact, each $\pi \in \widehat{G}$ is finite dimensional and $\pi(x) = (\pi_{ij}(x))$ is a matrix. By the Peter-Weyl Theorem, we have in $L^2(G)$,

$$f = \sum_{\pi \in \widehat{G}} \sum_{1 \le i, j \le \dim \pi} (\dim \pi)\widehat{f}(\pi)_{ji}\pi_{ij}$$

where $\widehat{f}(\pi)_{ij} = \int_G f(x)\pi_{ij}(x^{-1})d\lambda(x)$. We have $\widehat{f}(\pi)\widehat{\sigma}(\pi) = \widehat{f}(\pi)$ and by Lemma 2.1.4, the matrix $I_{H_\pi} - \widehat{\sigma}(\pi)$ is invertible for $\pi \in \widehat{G}\backslash\{\iota\}$ which gives $\widehat{f}(\pi) = 0$ for $\pi \ne \iota$. So we have

$$f = \widehat{f}(\iota) = \int_G f(x)d\lambda(x).$$

$\square$

The following result generalizes Lemma 2.1.6.

Proposition 2.1.7. *Let G be a central group and let σ be an adapted probability measure on G. If $f \in L^1(G)$ satisfies $\sigma * f = f$ and the function $\pi \mapsto (\dim \pi)tr(\widehat{f}(\pi)\pi(x)^*) \in L^1(\widehat{G})$ for $x \in G$, then f is constant.*

Proof. By Lemma 2.1.4, 1 is not an eigenvalue of the matrix $\widehat{\sigma}(\pi)$ for all $\pi \in \widehat{G}\backslash\{\iota\}$. So $\widehat{\sigma}(\pi) - I_{H_\pi}$ is invertible for such π. From Lemma 2.1.5, we have $\widehat{f}(\pi)(\widehat{\sigma}(\pi) - I_{H_\pi}) = 0$ which implies $\widehat{f}(\pi) = 0$ for all $\pi \in \widehat{G}\backslash\{\iota\}$. The inversion formula in (2.1) therefore reduces to $f(x) = \widehat{f}(\iota)\eta\{\iota\}$ for all $x \in G$.

$\square$

Remark 2.1.8. Let $\sigma \in M(G)$ and $\alpha \in \mathbb{C}$ with $|\alpha| > \|\sigma\| > 0$. If $f \in L^p(G)$ with $1 \leq p \leq \infty$ and $\sigma * f = \alpha f$, then $f = 0$. For if $\|f\| > 0$, then we have $\|f\| \leq \frac{\|\sigma\|\|f\|}{|\alpha|} < \|f\|$. Hence $J_\sigma^\perp = \{0\}$ if $\|\sigma\| < 1$.

We note that central groups have the Liouville property as shown in [28; Theorem 5.1]. We give an alternative proof below. The definition of a right uniformly continuous function is given in the next section.

Proposition 2.1.9. *Let σ be a non-degenerate probability measure on a central group G and let $f : G \to \mathbb{R}$ be a bounded right uniformly continuous σ-harmonic function. Then f is constant.*

Proof. By [14; Lemma 1.1], we have

$$f(az) = f(a) \quad (a \in G, z \in Z)$$

where Z is the centre of G. We can therefore define $\tilde{f} : G/Z \to \mathbb{R}$ by $\tilde{f}(aZ) = f(a)$ for $a \in G$. Let $\tilde{\sigma}$ be the image measure on G/Z of σ under the quotient map $q : G \to G/Z$. Then $\tilde{\sigma}$ is non-degenerate and $\tilde{\sigma} * \tilde{f} = \tilde{f}$. By compactness of G/Z, $\tilde{f}$, and hence f, is constant. $\square$

Example 2.1.10. We note that the Choquet-Deny Theorem does not hold for complex measures. Define $\sigma \in M^1(\mathbb{R})$ by

$$\sigma(S) = \frac{1}{2\pi} \int_{S \cap [0,2\pi]} e^{ix} dx$$

where S is a Borel subset of $\mathbb{R}$. Then $\operatorname{supp} |\sigma| = [0, 2\pi]$, $\sigma(\mathbb{R}) = 0$ and $\|\sigma\| = |\sigma|(\mathbb{R}) = \frac{1}{2\pi} \int_0^{2\pi} |e^{ix}| dx = 1$. Let $f : \mathbb{R} \longrightarrow \mathbb{C}$ be given by $f(x) = e^{ix}$. Then f is a nonconstant bounded σ-harmonic frunction and $\sigma * f^2 = 0$.

In the following section, we will derive a Poisson representation of bounded σ-harmonic functions for $\sigma \in M(G)$ with $\|\sigma\| = 1$.

2.2. Poisson representation of harmonic functions

In this section, we derive a Poisson representation for bounded uniformly continuous *complex* harmonic functions on a locally compact group. This is achieved by introducing an abelian C^*-algebraic structure on these functions which enables us to use the Gelfand transform to obtain the Poisson space for the representation. Such technique was first used by Furstenberg [27] for bounded

σ-harmonic functions on a semisimple Lie group, and extended to locally compact groups G by Azencott [5], where σ is an absolutely continuous probability measure and the C^*-product of two uniformly continuous harmonic functions f and h is given by the formula

$$(f \cdot h)(x) = \lim_{n \to \infty} \int_G f(y^{-1}x)h(y^{-1}x)d\sigma^n(y)$$

where σ^n is the n-times convolution of σ. For arbitrary locally compact group G and complex measure $\sigma \in M^1(G)$, the constant functions need not be σ-harmonic and the C^*-structure is obtained in a more elaborate way, using the structure theory of contractive projections on abelian C^*-algebras, but the C^*-structure thus obtained coincides with the one above in [5,27] if σ is a probability measure. Therefore our construction gives a unified approach to the Poisson representation as well as some functional analytic insights.

We have already mentioned that in the case of a probability measure σ, Paterson [63,64] has also obtained a Poisson space using a contractive projection, but the C^*-structure in this case is simpler as the constant-1 function is σ-harmonic. For complex measures σ, we use Friedman and Russo's result [26] to obtain the C^*-structure in terms of an extremal harmonic function.

Let $\sigma \in M(G)$ be absolutely continuous and let $f \in L^\infty(G)$. Then $\sigma * f$ is right uniformly continuous (cf. [38, 20.16]) and so is every bounded σ-harmonic function on G where a function $\varphi : G \to \mathbb{C}$ is *right uniformly continuous* if for any $\varepsilon > 0$, there exists a neighbourhood U of the identity such that $xy^{-1} \in U$ implies $|\varphi(x) - \varphi(y)| < \varepsilon$. If we define the *left* and *right* *translates* of φ through y by

$$(L_y\varphi)(x) = \varphi(y^{-1}x) = (\delta_y * \varphi)(x),$$

$$(R_y\varphi)(x) = \varphi(xy) = \frac{1}{\Delta(y)} (\varphi * \delta_{y^{-1}})(x)$$

then a function φ in $L^\infty(G)$ is right uniformly continuous if, and only if,

$$\|L_y\varphi - \varphi\|_\infty \to 0 \quad \text{as} \quad y \to e$$

where $\| \cdot \|_\infty$ denotes the essential sup norm, φ is *left uniformly continuous* if $\|R_y\varphi - \varphi\|_\infty \to 0$ as $y \to e$. Let $C_b(G)$ denote the C^*-algebra of bounded continuous functions on G and let $C_{\ell u}(G)$ $[C_{ru}(G)]$ be the C^*-subalgebra of $C_b(G)$ consisting of the left [right] uniformly continuous functions. We note that $C_{\ell u}(G) = C_{ru}(G)$ if, and only if, G is a [SIN]-group [61]. We introduce an algebraic product in the dual $C_{\ell u}(G)^*$ as follows. For each $m \in C_{\ell u}(G)^*$ and $f \in C_{\ell u}(G)$, we define $m \circ f \in L^\infty(G)$ by

$$\langle \varphi, m \circ f \rangle = \langle f * \check{\varphi}, m \rangle \qquad (2.2)$$

where $\varphi \in L^1(G)$ and $\check{\varphi}(x) = \varphi(x^{-1})$. Then we have

$$m \circ f \in C_{\ell u}(G) \quad \text{and} \quad (m \circ f)(x) = \langle R_x f, m \rangle$$

for $x \in G$ [54, Lemma 3]. Now, for $m, n \in C_{\ell u}(G)^*$, we define their product $m \circ n$ by

$$\langle f, m \circ n \rangle = \langle m \circ f, n \rangle \qquad (2.3)$$

where $f \in C_{\ell u}(G)$. Then $(C_{\ell u}(G)^*, \circ)$ is a Banach algebra and the product is weak*- continuous when the first variable is kept fixed. Given $a \in G$ and the point mass $\delta_a \in C_{\ell u}(G)^*$, we have $\delta_a \circ f = L_{a^{-1}} f$ and therefore $\delta_a \circ \delta_b = \delta_{ab}$ for any $b \in G$. There is a linear isometry $\mu \in M(G) \mapsto \tilde{\mu} \in C_{\ell u}(G)^*$ where $\tilde{\mu}$ is defined by

$$\langle f, \tilde{\mu} \rangle = \int_G f d\mu$$

for $f \in C_{\ell u}(G)$. We have $\widetilde{\mu * \nu} = \tilde{\mu} \circ \tilde{\nu}$ for $\mu, \nu \in M(G)$.

Let $\sigma \in M(G)$ and let J_σ be the norm-closure of $\{\check{\sigma} * f - f : f \in L^1(G)\}$, as defined in Section 2.1. Then J_σ is a closed right ideal of $L^1(G)$ and if σ is a probability measure, then J_σ is contained in the ideal

$$L^1_0(G) = \left\{ f \in L^1(G) : \int_G f d\lambda = 0 \right\}.$$

The annihilator $J_\sigma^\perp = (L^1(G)/J_\sigma)^*$ is the right-translation invariant subspace of $L^\infty(G)$ consisting of the (essentially) bounded σ-harmonic functions on G. If σ is a probability measure, then evidently $J_\sigma = L^1_0(G)$ if, and only if, $\dim J_\sigma^\perp = 1$, in other words; the bounded σ-harmonic functions are constant. The algebraic structure of $L^1(G)/J_\sigma$ has been studied in detail in [83] when σ is a probability measure.

Definition 2.2.1. Let $\sigma \in M(G)$. We denote by $H(\sigma)$ the closed subspace of $L^\infty(G)$ consisting of (left and right) uniformly continuous σ-harmonic functions on G.

We have $H(\sigma) \subset J_\sigma^\perp \cap C_{\ell u}(G)$ and they are equal if σ is absolutely continuous since, as remarked before, absolute continuity of σ implies that the σ-harmonic functions are right uniformly continuous.

Lemma 2.2.2. *Given* $\sigma \in M(G)$, *then* $J_\sigma^\perp \cap C_{\ell u}(G)$ *is weak*-dense in* $J_\sigma^\perp$.

Proof. Let $\{\varphi_\alpha\}$ be a bounded approximate identity in $L^1(G)$ and let $f \in J_\sigma^\perp$. Then $f * \check\varphi_\alpha \in J_\sigma^\perp \cap C_{\ell u}(G)$ and for $\psi \in L^1(G)$, we have

$$\langle \psi, f \rangle = \lim_\alpha \langle \psi * \varphi_\alpha, f \rangle = \lim_\alpha \langle \psi, f * \check\varphi_\alpha \rangle.$$

$\square$

To obtain a Poisson representation of $H(\sigma)$ we need to turn $J_\sigma^\perp$ into an abelian C^*-algebra, but for this we cannot use the pointwise product in $L^\infty(G)$ since $J_\sigma^\perp$ is not closed with respect to this product except in the trivial case.

Lemma 2.2.3. *Let* σ *be a probability measure on* G *and let* $h \in J_\sigma^\perp$ *be continuous such that* $h^2 \in J_\sigma^\perp$. *Then* h *is constant on the subgroup generated by* $\mathrm{supp}\ \sigma$.

Proof. It suffices to show that $h(sx) = h(x)$ for all $x \in G$ and $s \in \mathrm{supp}\ \sigma$. Let $x \in G$ and consider $h_x(\cdot) = h(\cdot x)$. Then $\sigma * h \in C_b(G)$. We have $h(x) = \int_G h(y^{-1}x)d\sigma(y) = \int_G h_x(y^{-1})d\sigma(y)$. Since $h^2 \in J_\sigma^\perp$, we have

$$h^2(x) = \int_G h_x^2(y^{-1})d\sigma(y) = \left(\int_G h_x(y^{-1})d\sigma(y)\right)^2$$

which forces $h_x(s^{-1}) = $ constant c, say, for all $s \in \mathrm{supp}\ \sigma$. Indeed, if $c^2 = \int_G h_x^2(y^{-1})d\sigma(y) \neq 0$, then $\int_G F(y)d\sigma(y) = 1$ where $F(y) = \frac{1}{c}h_x(y^{-1})$. As $\int_G F^2(y)d\sigma(y) + 1 = 2$ and $2F \leq F^2 + 1$, we have $2F = F^2 + 1$ a.e. $[\sigma]$ which gives $F = 1$ a.e. $[\sigma]$ and $h_x(s^{-1}) = c$ on $\mathrm{supp}\ \sigma$. It follows that $h(x) = \int_G c\,d\sigma = c = h(s^{-1}x)$ for all $s \in \mathrm{supp}\ \sigma$. $\square$

Corollary 2.2.4. *Let* σ *be an adapted probability measure on* G. *The following conditions are equivalent:*

(i) $H(\sigma)$ *is a subalgebra of* $L^\infty(G)$;

(ii) $H(\sigma) = \mathbb{C}\mathbf{1}$ *where* $\mathbf{1}$ *is the constant-1 function.*

To introduce a C^*-product in $H(\sigma) \subset J_\sigma^\perp$ for $\sigma \in M^1(G)$, usually different from the $L^\infty(G)$-product, we construct a *contractive projection*

$$P : L^\infty(G) \to J_\sigma^\perp$$

(that is, a surjective linear contraction satisfying $P^2 = P$) and use the structure theory of P, developed in [26,59], to give $J_\sigma^\perp$ an abelian C^*-algebraic structure. If σ is absolutely continuous, this yields readily a Poisson representation of $J_\sigma^\perp$ and hence of $H(\sigma) = J_\sigma^\perp \cap C_{\ell u}(G)$. We note that, however, $H(\sigma)$ need not be a subalgebra of $J_\sigma^\perp$ and that the group G may not act continuously on the Poisson space for the representation. But if the identity of $J_\sigma^\perp$ falls into $H(\sigma)$, then $H(\sigma)$ is a subalgebra of $J_\sigma^\perp$ and the group action is continuous. In particular, if σ is a probability measure, then the constant-1 function is the identity lying in $J_\sigma^\perp \cap C_{\ell u}(G)$ which becomes a C^*-subalgebra of $J_\sigma^\perp$ and our C^*-product coincides with the product defined in [5,27].

Let $\sigma \in M^1(G)$. We now construct a contractive projection $P : L^\infty(G) \longrightarrow J_\sigma^\perp$.

Proposition 2.2.5. *Let* $\sigma \in M^1(G)$. *Then there is a contractive projection* $P_\sigma : L^\infty(G) \to J_\sigma^\perp$ *satisfying* $P_\sigma(\sigma * f) = \sigma * P_\sigma(f)$ *for all* $f \in L^\infty(G)$.

Proof. Let $T_\sigma : L^\infty(G) \to L^\infty(G)$ be the convolution operator $T_\sigma(f) = \sigma * f$ and let $\mathcal{G}$ be the closed convex hull of $\{T_\sigma^n : n \geq 1\}$ in $L^\infty(G)^{L^\infty(G)}$ where $L^\infty(G)$ is equipped with the weak*-topology. Then for each $f \in L^\infty(G)$, the orbit $\mathcal{G}(f) = \{T(f) : T \in \mathcal{G}\}$ is a weak*-compact convex set in $L^\infty(G)$ and therefore $T_\sigma : \mathcal{G}(f) \to \mathcal{G}(f)$ has a fixed-point, that is, $\mathcal{G}(f) \cap J_\sigma^\perp \neq \emptyset$. By [55, Theorem 2.1], there exists $P_\sigma \in \mathcal{G}$ such that $P_\sigma(f) \in J_\sigma^\perp$ for $f \in L^\infty(G)$. This defines a map $P_\sigma : L^\infty(G) \to J_\sigma^\perp$ which is contractive since $\|\sigma\| = 1$. For $f \in L^\infty(G)$, we have $P_\sigma(\sigma * f) = \sigma * P_\sigma(f)$ since $P_\sigma \in \overline{co}\{T_\sigma^n : n \geq 1\}$. Given $h \in J_\sigma^\perp$, we have $T_\sigma^n(h) = h$ for all $n \geq 1$. Hence $P_\sigma(h) = h$, that is, P_σ is a projection onto $J_\sigma^\perp$. □

Remark 2.2.6. From the above construction of P_σ, we see that P_σ commutes with the right translations, that is, $P_\sigma(R_x f) = R_x P_\sigma(f)$ for all $x \in G$. Also, there is a net (μ_α) in the convex hull of $\{\sigma^n : n \geq 1\}$ such that $(\mu_\alpha * f)$ weak*-converges to $P_\sigma(f)$ for all $f \in L^\infty(G)$. By the Mackey-Arens Theorem [69], the net (μ_α) can be chosen such that $(\mu_\alpha * f)$ converges to $P_\sigma(f)$ uniformly on weakly compact sets in $L^1(G)$.

Remark 2.2.7. Let **1** be the constant-1 function in $L^\infty(G)$. Then $P_\sigma(\mathbf{1})$ is a constant function in $L^\infty(G)$ and since $\sigma * P_\sigma(\mathbf{1}) = P_\sigma(\mathbf{1})$, we have $P_\sigma(\mathbf{1}) = \mathbf{1}$ if $\sigma(G) = 1$; but $P_\sigma(\mathbf{1}) = 0$ if $\sigma(G) \neq 1$.

We note that Proposition 2.2.5 gives the following partial proof of Corollary 2.2.8 (see also [64, p.249]).

Corollary 2.2.8. *If there exists* $\sigma \in M_+^1(G)$ *such that* $J_\sigma^\perp = \mathbb{C}\mathbf{1}$, *then* G *is amenable.*

Proof. Let $P_\sigma : L^\infty(G) \longrightarrow J_\sigma^\perp = \mathbb{C}1$ be the contractive projection in Proposition 2.2.5. Then the functional $m : L^\infty(G) \longrightarrow \mathbb{C}$ defined by $P_\sigma(f) = m(f)1$ for $f \in L^\infty(G)$ is a right-invariant mean.

Let A be a unital abelian C^*-algebra and let $P : A \to A$ be a contractive projection. Lindenstrauss and Wulbert [59] have shown that the range $P(A)$ is linearly isometric to a unital abelian C^*-algebra if, and only if, the closed unit ball of $P(A)$ contains an extreme point. In [26], Friedman and Russo show that the result remains true without A being unital and they also construct an explicit C^*-structure on $P(A)$ as follows. Let u be an extreme point of the closed unit ball of $P(A)$. For $f, g \in P(A)$, define

$$f \underset{u}{\times} g = P(fu^*g)$$

$$f^\star = P(uf^*u).$$

Then $\left(P(A), \underset{u}{\times}, \star\right)$ is an abelian C^*-algebra with identity u and the original norm.

Corollary 2.2.9. *Let* $\sigma \in M^1(G)$. *Then* $J_\sigma^\perp$ *is linearly isometric to an abelian von Neumann algebra.*

Proof. This follows from the above remarks since $J_\sigma^\perp$ is a dual space and its closed unit ball contains extreme points.

The above corollary implies immediately that $L^1(G)/J_\sigma$ is isometrically isomorphic to some L^1-space which gives an alternative proof of part of [83, Theorem 2.1] as well as extending it to the case of complex σ.

We note that the extreme points of the closed unit ball of $L^\infty(G)$ are exactly the unitaries, that is, functions taking values in the unit circle $\mathbb{T}$. Hence a function $u : G \longrightarrow \mathbb{T}$ satisfying $\sigma * u = u$ is an extreme point of the closed unit ball of $J_\sigma^\perp$.

From now on, we fix an extreme point u of the closed unit ball of $J_\sigma^\perp$. Then $J_\sigma^\perp$ is an abelian von Neumann algebra with identity u, under the following product and involution:

$$f \underset{u}{\times} g = P_\sigma(f\bar{u}g)$$

$$f^* = P_\sigma(u\bar{f}u).$$

If σ is a probability measure, then $1 \in J_\sigma^\perp$ and we will take $u = 1$ in which case we have the product $f \times g = P_\sigma(fg)$ and involution $f^* = P_\sigma(\bar{f})$. By the Banach-Stone Theorem, the product $f \times g$ is the same as the one defined in [5,27] as the pointwise limit of the sequence $(\sigma^n * fg)$. We will show that $f \times g$ is also the weak*-limit of $(\sigma^n * fg)$. We first proceed to the Poisson representation of $J_\sigma^\perp$ with absolutely continuous σ.

Let $J_\sigma^\perp$ be equipped with the above abelian von Neumann algebraic structure in which the identity is u. Let Σ_σ be the pure state space of $J_\sigma^\perp$ so that $J_\sigma^\perp$ is isometrically $*$-algebraic isomorphic to $C(\Sigma_\sigma)$ via the Gelfand map $f \in J_\sigma^\perp \mapsto \hat{f} \in C(\Sigma_\sigma)$ where $\hat{f}(s) = s(f)$ for $s \in \Sigma_\sigma$. We also write $\langle f, s \rangle$ for $s(f)$ and note that Σ_σ is a Stonean space.

For $x \in G$, the right-translation $R_x : J_\sigma^\perp \to J_\sigma^\perp$ induces an isometry $\hat{f} \in C(\Sigma_\sigma) \mapsto R_x \cdot \hat{f} \in C(\Sigma_\sigma)$ where $R_x \cdot \hat{f} = \widehat{R_x f}$. By the Banach-Stone Theorem, there is a homeomorphism $\tau_x : \Sigma_\sigma \to \Sigma_\sigma$ such that

$$\langle R_x f, s \rangle = \widehat{R_x f}(s) = \widehat{R_x u}(s) \hat{f}(\tau_x(s))$$

$$= \langle R_x u, s \rangle \hat{f}(\tau_x(s)) \tag{2.4}$$

where $|\langle R_x u, s \rangle| = 1$ for all $s \in \Sigma_\sigma$.

Lemma 2.2.10. *For $x \in G$ and $s \in \Sigma_\sigma$, we have $\overline{\langle R_x u, s \rangle}\, R_x^*(s) \in \Sigma_\sigma$.*

Proof. It suffices to show that $\overline{\langle R_x u, s \rangle}\, R_x^*(s)$ is a multiplicative functional on $J_\sigma^\perp$. Let $f, g \in J_\sigma^\perp$. Then

$$\langle f \underset{u}{\times} g,\, \overline{\langle R_x u, s \rangle}\, R_x^*(s) \rangle = \overline{\langle R_x u, s \rangle}\, \langle R_x(f \underset{u}{\times} g), s \rangle$$

$$= \overline{\langle R_x u, s \rangle}\, \langle R_x u, s \rangle\, \widehat{f \underset{u}{\times} g}\,(\tau_x(s))$$

$$= \hat{f}(\tau_x(s))\, \hat{g}(\tau_x(s))$$

$$= \overline{\langle R_x u, s \rangle}\, \langle f, R_x^*(s) \rangle\, \overline{\langle R_x u, s \rangle}\, \langle g, R_x^*(s) \rangle$$

by (2.4). $\qquad\square$

Lemma 2.2.11. *The map $(x, s) \in G \times \Sigma_\sigma \mapsto x \cdot s \in \Sigma_\sigma$, defined by*

$$x \cdot s = \overline{\langle R_{x^{-1}} u, s \rangle}\, R_{x^{-1}}^*(s),$$

is a group action.

Proof. Let $x, y \in G$ and $s \in \Sigma_\sigma$. We show that $y \cdot (x \cdot s) = (yx) \cdot s$. Write $t = x \cdot s$. Then

$$y \cdot (x \cdot s) = \overline{\langle R_{y^{-1}} u, t \rangle} \, R^*_{y^{-1}}(t)$$

$$= \overline{\langle R_{y^{-1}} u, \overline{\langle R_{x^{-1}} u, s \rangle} \, R^*_{x^{-1}}(s) \rangle} \, R^*_{y^{-1}} \left(\overline{\langle R_{x^{-1}} u, s \rangle} \, R^*_{x^{-1}}(s) \right)$$

$$= \langle R_{x^{-1}} u, s \rangle \, \overline{\langle R_{y^{-1}} u, R^*_{x^{-1}}(s) \rangle} \, \overline{\langle R_{x^{-1}} u, s \rangle} \, R^*_{y^{-1}} R^*_{x^{-1}}(s)$$

$$= \overline{\langle R_{x^{-1}} R_{y^{-1}} u, s \rangle} \, R^*_{(yx)^{-1}}(s) = (yx) \cdot s.$$

$\square$

Notation 2.2.12. For $x \in G$ and $s \in \Sigma_\sigma$, we write $\langle R_x u, s \rangle = e^{i\theta(x,s)}$ where $\theta : G \times \Sigma_\sigma \to [0, 2\pi)$.

Now we have the Poisson representation of $J_\sigma^\perp$.

Proposition 2.2.13. *Let $\sigma \in M^1(G)$ be absolutely continuous. Then there exists a complex Borel measure $\tilde\nu_\sigma$ on Σ_σ such that for each $f \in J_\sigma^\perp$, we have*

$$f(x) = \int_{\Sigma_\sigma} \widehat{f}(x^{-1} \cdot s) e^{i\theta(x,s)} d\tilde\nu_\sigma(s) \quad (x \in G).$$

Proof. Since σ is absolutely continuous, we have $J_\sigma^\perp \subset C_{ru}(G)$ and we can define a measure $\tilde\nu_\sigma$ on Σ_σ by

$$\tilde\nu_\sigma(\widehat{f}) = f(e)$$

for $\widehat{f} \in C(\Sigma_\sigma)$. Then for $f \in J_\sigma^\perp$ and $x \in G$, we have, from (2.4),

$$f(x) = R_x f(e) = \tilde\nu_\sigma(\widehat{R_x f})$$

$$= \int_{\Sigma_\sigma} \widehat{R_x f}(s) d\tilde\nu_\sigma(s)$$

$$= \int_{\Sigma_\sigma} \overline{\langle R_x u, s \rangle} \, \langle f, R^*_x s \rangle \, \langle R_x u, s \rangle \, d\tilde\nu_\sigma(s)$$

$$= \int_{\Sigma_\sigma} \widehat{f}(x^{-1} \cdot s) e^{i\theta(x,s)} d\tilde\nu_\sigma(s).$$

$\square$

We now consider the space $J_\sigma^\perp \cap C_{\ell u}(G)$. Let $\omega : (J_\sigma^\perp)^* \to (J_\sigma^\perp \cap C_{\ell u}(G))^*$ be the restriction map. We define an equivalence relation $\sim$ on Σ_σ by

$$s \sim s' \iff \omega(s) = \omega(s') \quad \text{and} \quad \langle R_x u, s \rangle = \langle R_x u, s' \rangle \; \forall x \in G.$$

Let $\Pi_\sigma = \Sigma_\sigma / \sim$ be equipped with the quotient topology which is compact, and let ν_σ be the image measure on Π_σ of $\tilde{\nu}_\sigma$ by the quotient map $\Sigma_\sigma \to \Sigma_\sigma / \sim$. The group action $G \times \Sigma_\sigma \to \Sigma_\sigma$ defined in Lemma 2.2.11 induces a group action $G \times \Pi_\sigma \to \Pi_\sigma$ given by $x \cdot [s]_\sim = [x \cdot s]_\sim$ for $x \in G$ and $[s]_\sim \in \Pi_\sigma$. Given $f \in J_\sigma^\perp \cap C_{\ell u}(G)$ and $x \in G$, we regard $\widehat{R_x f}$ as a function on Π_σ by the following well-defined identification

$$\widehat{R_x f}([s]_\sim) = \langle R_x f, [s]_\sim \rangle = \langle R_x f, s \rangle.$$

Likewise we write

$$e^{i\theta(x,[s]_\sim)} = \langle R_x u, [s]_\sim \rangle = \langle R_x u, s \rangle$$

for $x \in G$ and $[s]_\sim \in \Pi_\sigma$.

Remark 2.2.14. If $u \in J_\sigma^\perp \cap C_{\ell u}(G)$, then Π_σ identifies with $\omega(\Sigma_\sigma)$ and the above group action can be written as $x \cdot t = \overline{\langle R_{x^{-1}} u, t \rangle}\, R_{x^{-1}}^*(t)$ for $(x, t) \in G \times \Pi_\sigma$.

Proposition 2.2.15. *Let $\sigma \in M^1(G)$ be absolutely continuous. Then there exists a complex Borel measure ν_σ on Π_σ and a function $\theta : G \times \Pi_\sigma \to [0, 2\pi)$ such that for each $f \in H(\sigma)$, we have*

$$f(x) = \int_{\Pi_\sigma} \hat{f}(x^{-1} \cdot t) e^{i\theta(x,t)} d\nu_\sigma(t) \quad (x \in G).$$

Proof. Let $f \in H(\sigma)$ and let ν_σ be defined as above. Then, by Proposition 2.2.13, we have

$$f(x) = \int_{\Sigma_\sigma} \langle R_x f, s \rangle \, d\tilde{\nu}_\sigma(s)$$

$$= \int_{\Pi_\sigma} \langle R_x f, [s]_\sim \rangle \, d\nu_\sigma([s]_\sim)$$

$$= \int_{\Pi_\sigma} \overline{\langle R_x u, [s]_\sim \rangle} \, \langle R_x f, [s]_\sim \rangle \, \langle R_x u, [s]_\sim \rangle \, d\nu_\sigma([s]_\sim)$$

$$= \int_{\Pi_\sigma} \widehat{f}(x^{-1} \cdot [s]_\sim) e^{i\theta(x, [s]_\sim)} \, d\nu_\sigma([s]_\sim)$$

for $x \in G$. $\qquad\qquad\qquad\qquad\qquad\qquad\qquad\qquad\qquad\qquad$ □

Although the spaces Σ_σ, Π_σ and the group action are constructed in terms of the chosen identity $u \in J_\sigma^\perp$, they are unique in the following sense. Let $v \in J_\sigma^\perp$ be another extreme point of the unit ball such that $(J_\sigma^\perp, \times, \underset{v}{*}, \| \cdot \|_\infty)$ is an abelian C^*-algebra isomorphic to $C(\Sigma_\sigma^v)$ via the Gelfand map where $\Sigma_\sigma^v \subset (J_\sigma^\perp)^*$. Then the identity map from $(J_\sigma^\perp, \underset{u}{\times}, \underset{u}{*})$ to $(J_\sigma^\perp, \underset{v}{\times}, \underset{v}{*})$ is an isometry and hence there is a homeomorphism $\tau : \Sigma_\sigma^v \to \Sigma_\sigma$ such that, for $f \in J_\sigma^\perp$ with $\widehat{f} \in C(\Sigma_\sigma)$, we have

$$\widehat{f}(\tau(t)) = \overline{\langle u, t \rangle} \, \langle f, t \rangle \quad (t \in \Sigma_\sigma^v).$$

Further, the homeomorphism is *equivariant*, that is, for $x \in G$, we have $\tau(x \cdot t) = x \cdot \tau(t)$ where

$$x \cdot t = \overline{\langle R_{x^{-1}} v, t \rangle} \, R_{x^{-1}}^*(t)$$

and $\qquad x \cdot \tau(t) = \overline{\langle R_{x^{-1}} u, \tau(t) \rangle} \, R_{x^{-1}}^*(\tau(t)).$

If u and v are in $J_\sigma^\perp \cap C_{\ell u}(G)$, then Π_σ and Π_σ^v are equivariantly homeomorphic as above.

Definition 2.2.16. Given an absolutely continuous measure $\sigma \in M^1(G)$, we call Π_σ and ν_σ, constructed above, the *Poisson space* and the *Poisson measure* for σ, *with respect to* u. The reference to u is often understood and omitted. If σ is a probability measure, we will always choose $u = \mathbf{1}$.

The Poisson representation points to an interesting and relevant object of study, namely, the Poisson space Π_σ which should reveal useful structural information about $H(\sigma)$ and $J_\sigma^\perp$. We will show some interesting properties of Π_σ and $H(\sigma)$ in the case of the extreme point $u \in J_\sigma^\perp$ falling into $H(\sigma)$ which occurs, for instance, when σ is a probability measure and u chosen to be the constant-1 function in $H(\sigma)$. We note that, for absolutely continuous σ, $J_\sigma^\perp = H(\sigma)$ if G is a [SIN]-group and in this case, Π_σ is a Stonean space by Corollary 2.2.9.

Assuming and fixing $u \in J_\sigma^\perp \cap C_{\ell u}(G)$ for the rest of this section, we first investigate the C^*-structure of $J_\sigma^\perp \cap C_{\ell u}(G)$.

Theorem 2.2.17. *Let $\sigma \in M^1(G)$ and $u \in J_\sigma^\perp \cap C_{\ell u}(G)$. Then $J_\sigma^\perp \cap C_{\ell u}(G)$ is an abelian C^*-algebra and there is a net $\{\mu_\alpha\}$ in the convex hull $\mathrm{co}\{\sigma^n : n \geq 1\}$ such that the C^*-product is given by*

$$(f \underset{u}{\times} g)(x) = \lim_\alpha \int_G f(y^{-1}x)\bar{u}(y^{-1}x)g(y^{-1}x)d\mu_\alpha(y)$$

for $f, g \in J_\sigma^\perp \cap C_{\ell u}(G)$ and $x \in G$. In fact, $\mu_\alpha * (f\bar{u}g) \to f \underset{u}{\times} g$ uniformly on compact subsets of G.

Proof. Let $f, g \in J_\sigma^\perp \cap C_{\ell u}(G)$. Since $u \in C_{\ell u}(G)$ and P_σ commutes with right translations by Remark 2.2.6, we see that $f \underset{u}{\times} g = P_\sigma(f\bar{u}g)$ and $f^* = P_\sigma(u\bar{f}u)$ are in $C_{\ell u}(G)$. Hence $J_\sigma^\perp \cap C_{\ell u}(G)$ is a C^*-subalgebra of $J_\sigma^\perp$. Write $h = f\bar{u}g$ and let $K_h = \{\sum_{i=1}^n \lambda_i L_{x_i} h : x_i \in G, \sum_{i=1}^n |\lambda_i| \leq 1\}$. Denote by $\overline{K}_h$ the pointwise closure of K_h. We show that

$$\overline{K}_h = \{\mu \circ h : \mu \in C_b(G)^* \quad \text{and} \quad \|\mu\| \leq 1\} \subset C_{\ell u}(G)$$

where $\mu \circ h$ is defined in (2.2) and $\mu \circ h \in C_{\ell u}(G)$ with $(\mu \circ h)(x) = \langle R_x h, \mu \rangle$ for $x \in G$. Let $k \in \overline{K}_h$ be the pointwise limit of a net $k_\alpha = \sum_i \lambda_i^\alpha L_{x_i^\alpha} h$ in K_h. Let $\mu_\alpha = \sum_i \lambda_i^\alpha \delta_{(x_i^\alpha)^{-1}}$. Then we have $\mu_\alpha \in C_b(G)^*$ and $\|\mu_\alpha\| \leq 1$. Passing to a subnet, we may assume that (μ_α) weak*-converges to some $\mu \in C_b(G)^*$ with $\|\mu\| \leq 1$. Then for $y \in G$, we have

$$(\mu \circ h)(y) = \lim_\alpha (\mu_\alpha \circ h)(y) = \lim_\alpha \langle R_y h, \mu_\alpha \rangle = \lim_\alpha k_\alpha(y) = k(y).$$

Therefore $k = \mu \circ h$. Conversely, given $\mu \in C_b(G)^*$ with $\|\mu\| \leq 1$, it is the weak*-limit of a net (μ_α) of the form $\mu_\alpha = \sum_i \lambda_i^\alpha \delta_{x_i^\alpha}$ with $\sum_i |\lambda_i^\alpha| \leq 1$. As before, $\mu \circ h$ is the pointwise limit of $\mu_\alpha \circ h \in K_h$.

Next we show that $\overline{K}_h$ is equicontinuous. Let $\varepsilon > 0$. Then there is a neighbourhood V of e such that $v^{-1}u \in V$ implies $\|R_u h - R_v h\| < \varepsilon$ which gives $|(\mu \circ h)(v) - (\mu \circ h)(u)| \leq \|R_u h - R_v h\| < \varepsilon$ for all $\mu \in C_b(G)^*$ with $\|\mu\| \leq 1$.

By Remark 2.2.6, there is a net (μ_α) in $\mathrm{co}\{\sigma^n : n \geq 1\}$ such that $\mu_\alpha * h$ weak*-converges to $P_\sigma(h) = f \underset{u}{\times} g$. We have $\mu_\alpha * h = \check{\mu}_\alpha \circ h \in \overline{K}_h$. We can regard $\mu_\alpha \in C_b(G)^*$, and by taking a subnet, assume that $(\check{\mu}_\alpha)$ weak*-converges to some $\check{\mu} \in C_b(G)^*$ with $\|\check{\mu}\| \leq 1$. Then $(\check{\mu}_\alpha * h)$ is a net in $\overline{K}_h$ converging pointwise to $\check{\mu} \circ h \in \overline{K}_h$. By Ascoli's theorem, $(\mu_\alpha * h)$ converges to $\check{\mu} \circ h$ on compact subsets of G. So

$$\langle f, \mu_\alpha * h \rangle \to \langle f, \check{\mu} \circ h \rangle$$

for continuous functions f on G with compact support which are norm-dense in $L^1(G)$. As $\overline{K}_h$ is bounded by $\|h\|_\infty$ in $L^\infty(G)$, we have $(\mu_\alpha * h)$ converges to $\check{\mu} \circ h$ in the weak*-topology of $L^\infty(G)$. Hence $f \underset{u}{\times} g = \check{\mu} \circ h$ which completes the proof. $\qquad\square$

Corollary 2.2.18. *Let σ be an absolutely continuous probability measure on G. Then the C^*-product in $H(\sigma)$ is given by $f \times g = \mathrm{weak}^*\text{-}\lim\limits_{n \to \infty} \sigma^n * fg$ for $f, g \in H(\sigma)$.*

Proof. Since σ is a real measure, we have the involution $f^* = P_\sigma(\bar{f}) = \overline{P_\sigma(f)} = \bar{f}$ for $f \in H(\sigma)$. By [5,27], $H(\sigma)$ is an abelian C^*-algebra with product $(f \cdot g)(x) = \lim\limits_{n \to \infty} (\sigma^n * fg)(x)$ for $x \in G$. By the Banach-Stone Theorem, we have $f \times g = P_\sigma(fg) = f \cdot g$. As in the above proof, $(\sigma^n * fg)$ converges pointwise, and hence weak*, to $f \times g \in \overline{K}_{fg}$ by equicontinuity. $\qquad\square$

Now we investigate the action of G on the Poisson space Π_σ. For each $x \in G$, the homeomorphism $\tau_x : \Sigma_\sigma \to \Sigma_\sigma$ in (2.4) induces a weak*-continuous algebra homomorphism $\tilde{\tau}_x : J_\sigma^\perp \to J_\sigma^\perp$ given by

$$\langle \tilde{\tau}_x f, s \rangle = \overline{\langle R_x u, s \rangle} \, \langle R_x f, s \rangle \quad (f \in J_\sigma^\perp, s \in \Sigma_\sigma) \tag{2.5}$$

which is a *'perturbation'* of the right translation R_x by $\overline{R_x u}$. Given $u \in C_{\ell u}(G)$ and $f \in J_\sigma^\perp \cap C_{\ell u}(G)$, we have $\tilde{\tau}_x(f) \in J_\sigma^\perp \cap C_{\ell u}(G)$. We also note that, for $f \in J_\sigma^\perp$, we have $f \in C_{\ell u}(G)$ if, and only if, the map $x \in G \mapsto \tilde{\tau}_x f \in J_\sigma^\perp$ is continuous with respect to the norm-topology of $J_\sigma^\perp$.

Proposition 2.2.19. *Let $\sigma \in M^1(G)$ be absolutely continuous and $u \in H(\sigma)$. Then the group action*

$$(x, s) \in G \times \Pi_\sigma \mapsto \overline{\langle R_{x^{-1}} u, s \rangle} \, R_{x^{-1}}^*(s) \in \Pi_\sigma$$

is jointly continuous. If the group action $G \times \Sigma_\sigma \to \Sigma_\sigma$ is separately continuous, then $J_\sigma^\perp \subset C_{\ell u}(G)$, that is, $J_\sigma^\perp = H(\sigma)$.

Proof. For $u, f \in C_{\ell u}(G)$, we have $\|R_x u - u\| \to 0$ and $\|R_x f - f\| \to 0$ as $x \to e$. It follows that the group action $G \times \Pi_\sigma \to \Pi_\sigma$ is separately continuous, and hence jointly continuous by compactness of Π_σ [23].

Now if the action $G \times \Sigma_\sigma \to \Sigma_\sigma$ is separately continuous, then it is jointly continuous. Let $f \in J_\sigma^\perp$. To show $f \in C_{\ell u}(G)$, we show that the map $x \in G \mapsto \tilde{\tau}_x f \in J_\sigma^\perp$ is continuous. Let (x_α) be a net converging to x. Suppose $\|\tilde{\tau}_{x_\alpha} f - \tilde{\tau}_x f\| \not\to 0$. Then there is a subnet (x_β) such that

$$\|\tilde{\tau}_{x_\beta} f - \tilde{\tau}_x f\| \geq \epsilon > 0 \quad \text{for all} \quad \beta.$$

Let $s_\beta \in \Sigma_\sigma$ be such that $\|\tilde{\tau}_{x_\beta} f - \tilde{\tau}_x f\| = |\langle \tilde{\tau}_{x_\beta} f - \tilde{\tau}_x f, s_\beta \rangle|$. There is a subnet (s_γ) of (s_β) weak*-converging to some $s \in \Sigma_\sigma$. The jointly continuous group action gives the following contradiction:

$$0 < \epsilon \leq |\overline{\langle R_{x_\gamma} u, s \rangle} \langle R_{x_\gamma} f, s_\gamma \rangle - \overline{\langle R_x u, s_\gamma \rangle} \langle R_x f, s_\gamma \rangle|$$

$$\longrightarrow |\overline{\langle R_x u, s \rangle} \langle R_x f, s \rangle - \overline{\langle R_x u, s \rangle} \langle R_x f, s \rangle| = 0.$$

So $\|\tilde{\tau}_{x_\alpha} f - \tilde{\tau}_x f\| \to 0$ and $f \in C_{\ell u}(G)$. $\square$

Example. In general $J_\sigma^\perp \neq J_\sigma^\perp \cap C_{\ell u}(G)$. Let $\sigma \in M^1(G)$ be the natural extension of the Haar measure on a compact normal subgroup H of G. Then $J_\sigma^\perp = L^\infty(G/H)$ as the σ-harmonic functions are constant on the cosets of H. Since $J_\sigma^\perp \cap C_{\ell u}(G) = C_{\ell u}(G/H)$, we have $J_\sigma^\perp = J_\sigma^\perp \cap C_{\ell u}(G)$ if, and only if, G/H is discrete or equivalently, H is open.

Given $u \in J_\sigma^\perp \cap C_{\ell u}(G)$, then $J_\sigma^\perp \cap C_{\ell u}(G)$ is a C^*-subalgebra of $J_\sigma^\perp$ and $\Pi_\sigma = \omega(\Sigma_\sigma)$ is the pure state space of $J_\sigma^\perp \cap C_{\ell u}(G)$ which can be seen from the fact that the pure states of an abelian C^*-algebra are exactly the multiplicative states. We end this section with the following description of the state space $S(J_\sigma^\perp \cap C_{\ell u}(G))$ of $J_\sigma^\perp \cap C_{\ell u}(G)$.

For $x \in G$, we define $\bar{u}\delta_x \in C_{\ell u}(G)^*$ by

$$\langle f, \bar{u}\delta_x \rangle = \langle \bar{u}f, \delta_x \rangle = \bar{u}(x)f(x)$$

for $f \in C_{\ell u}(G)$. We also write $\bar{u}\delta_x$ for its restriction to $J_\sigma^\perp \cap C_{\ell u}(G)$.

Proposition 2.2.20. *Let $\sigma \in M^1(G)$ and $u \in J_\sigma^\perp \cap C_{\ell u}(G)$. Then $S(J_\sigma^\perp \cap C_{\ell u}(G)) = \overline{co}\{\bar{u}\delta_x : x \in G\}$ and $\Pi_\sigma \subset \{\bar{u}\delta_x : x \in G\}^-$ where the first and last bar '$-$' denote the weak*-closure and 'co' denotes the convex hull.*

Proof. We have $\{\bar{u}\delta_x : x \in G\} \subset S(J_\sigma^\perp \cap C_{\ell u}(G))$ since $(\bar{u}\delta_x)(u) = 1 = \|\bar{u}\delta_x\|$ and u is the identity of the C^*-algebra $J_\sigma^\perp \cap C_{\ell u}(G)$. Conversely, let $s \in S(J_\sigma^\perp \cap C_{\ell u}(G))$ and let $\tilde{s} \in C_{\ell u}(G)^*$ be a norm-preserving extension of s. Define $u\tilde{s} \in C_{\ell u}(G)^*$ by

$$(u\tilde{s})(f) = \tilde{s}(uf)$$

for $f \in C_{\ell u}(G)$. Then $(u\tilde{s})(1) = s(u) = 1 = \|u\tilde{s}\|$, that is, $u\tilde{s}$ is a state of $C_{\ell u}(G)$ and hence $u\tilde{s} \in \overline{co}\{\delta_x : x \in G\}$. Therefore $\tilde{s} \in \overline{co}\{\bar{u}\delta_x : x \in G\}$. The second assertion follows from Milman's theorem. $\quad\square$

We note that the intersection $\Pi_\sigma \cap \{\bar{u}\delta_x : x \in G\}$ is most likely empty as the following result shows.

Proposition 2.2.21. *Let $\sigma \in M^1(G)$ and $u \in J_\sigma^\perp \cap C_{\ell u}(G)$. Then $\Pi_\sigma \cap \{\bar{u}\delta_x : x \in G\} \neq \emptyset$ if, and only if, $f \underset{u}{\times} g = f\bar{u}g$ for all $f, g \in J_\sigma^\perp \cap C_{\ell u}(G)$, in which case $\Pi_\sigma = \{\bar{u}\delta_x : x \in G\}^-$.*

Proof. Let $s = \bar{u}\delta_x \in \Pi_\sigma$ for some $x \in G$. Then for any $y \in G$, a direct computation gives $\bar{u}\delta_{xy^{-1}} = y \cdot s \in \Pi_\sigma$. Hence $\{\bar{u}\delta_z : z \in G\}^- \subset \Pi_\sigma$ and they are therefore equal. Let $f, g \in H(\sigma)$ and consider $f \underset{u}{\times} g$, $f\bar{u}g \in C_{\ell u}(G)$. For every $z \in G$, we have $\langle f \underset{u}{\times} g, \bar{u}\delta_z \rangle = \langle f, \bar{u}\delta_z \rangle \langle g, \bar{u}\delta_z \rangle = \bar{u}(z)f(z)\bar{u}(z)g(z)$ and so $\delta_z(f \underset{u}{\times} g) = f(z)\bar{u}(z)g(z) = \delta_z(f\bar{u}g)$. Therefore $f \underset{u}{\times} g = f\bar{u}g$.

Conversely, the condition $f \underset{u}{\times} g = f\bar{u}g$ implies that $\bar{u}\delta_x$ is multiplicative on the C^*-algebra $J_\sigma^\perp \cap C_{\ell u}(G)$ for $x \in G$, that is, $\bar{u}\delta_x \in \Pi_\sigma$. $\quad\square$

If σ is a probability measure with $u = 1$, then the involution in $J_\sigma^\perp \cap C_{\ell u}(G)$ is the complex conjugation and the above result can be stated as follows.

Corollary 2.2.22. *Let σ be a probability measure on G. Then $\Pi_\sigma \cap \{\delta_x : x \in G\} \neq \emptyset$ if, and only if, $J_\sigma^\perp \cap C_{\ell u}(G)$ is a C^*-subalgebra of $L^\infty(G)$.*

2.3. Semigroup structures of the Poisson space

Given $\sigma \in M^1(G)$, the space $J_\sigma^\perp$ of bounded σ-harmonic functions is right-translation invariant. We first examine the case when $J_\sigma^\perp$ is also left-translation invariant. The motivation for such consideration is that, in this case, the Poisson space Π_σ admits a semigroup structure which is the main object of study in this section. We note that Paterson [64] has also studied a certain semigroup structure of Π_σ in the case when G is a compact topological semigroup and σ is a non-degenerate probability measure.

We observe that $J_\sigma^\perp$ is left-translation invariant if, and only if, $J_\sigma^\perp \cap C_{\ell u}(G)$ is so. This follows from the fact that $J_\sigma^\perp \cap C_{\ell u}(G)$ is weak*-dense in $J_\sigma^\perp$ and that left translation is weak*-continuous. Evidently, if σ is central which means that $\int_G h(xy)d\sigma(y) = \int_G h(yx)d\sigma(y)$ for all $h \in L^\infty(G)$, then $J_\sigma^\perp$ is translation invariant. The following lemma is easily verified.

Lemma 2.3.1. *Let $\sigma \in M(G)$. The following conditions are equivalent:*

(i) *$J_\sigma^\perp$ is translation invariant;*
(ii) *J_σ is a two-sided ideal in $L^1(G)$;*
(iii) *For every $h \in J_\sigma^\perp \cap C_{\ell u}(G)$, we have*

$$\int_G h(y^{-1}x)d\sigma(y) = \int_G h(xy^{-1})d\sigma(y).$$

Given that $J_\sigma^\perp$ is also left-translation invariant, the quotient $L^1(G)/J_\sigma$ is a Banach algebra and the dual $(J_\sigma^\perp)^* = (L^1(G)/J_\sigma)^{**}$ is also a Banach algebra with the *second* Arens product $\circ$ defined as follows:

(i) For $f \in J_\sigma^\perp$ and $p = \psi + J_\sigma \in L^1(G)/J_\sigma$, define $p \cdot f \in J_\sigma^\perp$ by

$$\langle q, p \cdot f \rangle = \langle \phi * \psi, f \rangle$$

where $q = \phi + J_\sigma \in L^1(G)/J_\sigma$.

(ii) For $f \in J_\sigma^\perp$ and $m \in (J_\sigma^\perp)^*$, define $m \circ f \in J_\sigma^\perp$ by

$$\langle p, m \circ f \rangle = \langle p \cdot f, m \rangle$$

for $p \in L^1(G)/J_\sigma$.

(iii) For $m, n \in (J_\sigma^\perp)^*$, define $m \circ n \in (J_\sigma^\perp)^*$ by

$$\langle f, m \circ n \rangle = \langle m \circ f, n \rangle$$

for $f \in J_\sigma^\perp$.

We note that, in the above definition, $p \cdot f = f * \check{\psi} \in J_\sigma^\perp \cap C_{\ell u}(G)$ and that $m \circ f \in J_\sigma^\perp \cap C_{\ell u}(G)$ for $f \in J_\sigma^\perp \cap C_{\ell u}(G)$. We also have $(m \circ f)(x) = \langle R_x f, m \rangle$ for $x \in G$ [54, Lemma 3]. It follows that the restriction map $\omega : (J_\sigma^\perp)^* \to (J_\sigma^\perp \cap C_{\ell u}(G))^*$ is an algebra homomorphism where $(J_\sigma^\perp \cap C_{\ell u}(G))^*$ is equipped with the Banach algebra product as defined in (2.3).

Lemma 2.3.2. *Let* u *be the identity in* $J_\sigma^\perp$. *The following conditions are equivalent:*

(i) u *is a multiplicative functional on* $L^1(G)/J_\sigma$;

(ii) u *is a continuous character on* G;

(iii) $u^{-1}(0) = \{\varphi + J_\sigma \in L^1(G)/J_\sigma : \langle \varphi, u \rangle = 0\}$ *is a two-sided ideal in* $L^1(G)/J_\sigma$;

(iv) $\{\varphi + J_\sigma \in L^1(G)/J_\sigma : \langle \varphi, u \rangle = 1$ *and* $\varphi + J_\sigma \geq 0\}$ *is a semigroup in* $L^1(G)/J_\sigma$ *where* $\varphi + J_\sigma \geq 0$ *means that* $\langle \varphi, f \rangle \geq 0$ *for every* $f \in J_\sigma^\perp$ *with* $f \geq 0$.

Proof. (i) $\Longrightarrow$ (ii). Lift u to a multiplicative linear functional u' on $L^1(G)$ via the quotient map. For $\varphi, \psi \in L^1(G)$, we have

$$\int_G u'(\varphi)\psi(y)u(y)d\lambda(y) = u'(\varphi)u'(\psi) = u'(\varphi * \psi)$$

$$= \int_G (\varphi * \psi)(x)u(x)d\lambda(x)$$

$$= \iint_G \Delta(y^{-1})\varphi(xy^{-1})\psi(y)u(x)d\lambda(x)d\lambda(y)$$

$$= \int_G u'(\Delta(y^{-1})R_{y^{-1}}\varphi)\psi(y)d\lambda(y)$$

which gives

$$u'(\varphi)u(y) = u'(\Delta(y^{-1})R_{y^{-1}}\varphi) \tag{2.6}$$

for λ-almost all $y \in G$ and that u is continuous λ-almost everywhere on G since the map $y \in G \mapsto \Delta(y^{-1})R_{y^{-1}}\varphi \in L^1(G)$ is continuous. We may therefore assume that u is continuous everywhere and (2.6) holds for all $y \in G$.

It follows that

$$u'(\varphi)u(xy) = u'\big(\Delta(y^{-1}x^{-1})R_{y^{-1}x^{-1}}\varphi\big)$$

$$= u'(\varphi)u(x)u(y)$$

for all $x, y \in G$ and u is a character on G as $\|u\|_\infty = 1$.

(i) $\implies$ (iv). If $\langle\varphi, u\rangle = \langle\psi, u\rangle = 1$ for $\varphi, \psi \in L^1(G)$, then $\langle(\varphi + J_\sigma) * (\psi + J_\sigma), u\rangle = \langle\varphi, u\rangle \langle\psi, u\rangle = 1 = \|(\varphi + J_\sigma) * (\psi + J_\sigma)\|$.

(iv) $\implies$ (i). Since u is multiplicative on the semigroup which generates $L^1(G)/J_\sigma$, u is multiplicative on $L^1(G)/J_\sigma$. $\qquad\qquad\square$

We note that $J_\sigma^\perp$ contains a character of G if, and only if, σ is of the form $u.\sigma_1$ where u is a character and σ_1 is a probability measure. Indeed, let $u \in J_\sigma^\perp$ be a character and let $d\sigma_1(x) = u(x^{-1})d\sigma(x)$. Then σ_1 is a probability measure since $\int_G u(x^{-1})d\sigma(x) = \sigma * u(e) = u(e) = 1$. Conversely, if σ is of the above form, then $\sigma * u = u$ by simple calculation.

We assume in the remaining section that $\sigma \in M^1(G)$ and $J_\sigma^\perp$ is translation invariant with identity u which is a continuous character on G.

We will show that Π_σ has a natural semigroup structure under the above assumption. We first recall some basic definitions.

A semigroup S is called a *left zero semigroup* if all of its elements are left zeros which means that $xy = x$ for all $x, y \in S$. Similarly S is called a *right zero semigroup* if $xy = y$ for all $x, y \in S$. By a *left [right] group* we mean a semigroup which is (isomorphic to) a direct product of a group and a right [left] zero semigroup. The (possibly empty) set of idempotents of a semigroup S is denoted by $E(S)$.

Let X, Y be nonempty sets and G be a group. Let $K = X \times G \times Y$. Given a map $\delta : X \times Y \to G$, we define a sandwich product on K by

$$(x, g, y) \circ (x', g', y') = (x, g\delta(y, x')g', y').$$

Then $(K, \circ)$ is a simple semigroup (i.e. K has no non-trivial two-sided ideal) and any semigroup isomorphic to a simple semigroup of this kind is called a *paragroup*.

Let S be a compact semigroup. It is called a *left topological semigroup* if the translations $x \mapsto sx$ $(s \in S)$ are continuous. S is called a *semitopological semigroup* if its multiplication is separately continuous. It is called a *topological semigroup* if the multiplication is jointly continuous.

Now we are ready to show some semigroup properties of the Poisson space Π_σ. Let

$$\mathcal{D}_\sigma = \{\overline{u}\,\delta_x : x \in G\}^-.$$

We have shown in Proposition 2.2.20 that $\Pi_\sigma \subset \mathcal{D}_\sigma$.

We note that $\mathcal{D}_\sigma \subset \left(J_\sigma^\perp \cap C_{\ell u}(G)\right)^*$ and the latter is a Banach algebra with the product $\circ$ defined in (2.3). We equip $\mathcal{D}_\sigma$ with the weak*-topology.

Theorem 2.3.3. *Let $J_\sigma^\perp$ be translation invariant with identity u which is a continuous character on G. Then $(\mathcal{D}_\sigma, \circ)$ is a compact left topological semigroup and $(\Pi_\sigma, \circ)$ is a closed subsemigroup of $\mathcal{D}_\sigma$ with idempotents. Further, the following conditions hold:*

(i) *Π_σ has a minimal ideal K and*

$$K \simeq E(p\Pi_\sigma) \times p\Pi_\sigma p \times E(\Pi_\sigma p)$$

where p is any idempotent of K and $p\Pi_\sigma = \{p \circ s : s \in \Pi_\sigma\}$ with similar definition for $p\Pi_\sigma p$ and $\Pi_\sigma p$. Also, $E(p\Pi_\sigma)$ is a right zero semigroup, $E(\Pi_\sigma p)$ is a left zero semigroup and $p\Pi_\sigma p = p\Pi_\sigma \cap \Pi_\sigma p$ is a group.

(ii) *The minimal ideal K need not be a direct product, but is a paragroup with respect to the natural map*

$$\delta : E(p\Pi_\sigma) \times E(\Pi_\sigma p) \to p\Pi_\sigma p : (x, y) \mapsto x \circ y.$$

(iii) *For any idempotent $p \in K$, $p\Pi_\sigma$ is a minimal right ideal and $\Pi_\sigma p$ is a minimal left ideal.*

(iv) *The minimal right ideals in Π_σ are closed and homeomorphic to each other.*

Proof. Since u is multiplicative, we have for $x, y \in G$,

$$(\overline{u}\,\delta_x) \circ (\overline{u}\,\delta_y) = (\overline{u}(x)\delta_x) \circ (\overline{u}(y)\delta_y) = \overline{u}(xy)\delta_{xy}$$

by (2.3). Let $m \in \mathcal{D}_\sigma$ and $a \in G$, with $m = w^* - \lim_\alpha \overline{u}\,\delta_{x_\alpha}$. Then for $f \in J_\sigma^\perp \cap C_{\ell u}(G)$,

$$\lim_\alpha \left\langle f, (\overline{u}\,\delta_{x_\alpha}) \circ (\overline{u}\,\delta_a) \right\rangle = \lim_\alpha \left\langle \overline{u}(a)R_a f, \overline{u}\,\delta_{x_\alpha} \right\rangle$$

$$= \langle \overline{u}(a)R_a f, m \rangle = \langle f, m \circ \overline{u}\,\delta_a \rangle.$$

So $m \circ \overline{u} \delta_a \in \mathcal{D}_\sigma$. Let $n \in \mathcal{D}_\sigma$ with $n = w^* - \lim_\beta \overline{u} \delta_{x_\beta}$. Then for $f \in J_\sigma^\perp \cap C_{\ell u}(G)$, we have

$$\langle f, m \circ n \rangle = \langle m \circ f, n \rangle$$
$$= \lim_\beta \left\langle m \circ f, \overline{u} \delta_{x_\beta} \right\rangle$$
$$= \lim_\beta \left\langle f, m \circ \overline{u} \delta_{x_\beta} \right\rangle$$

since $m \circ f \in J_\sigma^\perp \cap C_{\ell u}(G)$. Hence $m \circ n = w^* - \lim_\beta m \circ \overline{u} \delta_{x_\beta} \in \mathcal{D}_\sigma$ and $(\mathcal{D}_\sigma, \circ)$ is a compact left topological semigroup.

To see that Π_σ is a subsemigroup of $\mathcal{D}_\sigma$, we need to show that $s, s' \in \Pi_\sigma$ implies $s \circ s' \in \Pi_\sigma$. For this, it suffices to show that $s \circ s'$ is multiplicative on the abelian C^*-algebra $J_\sigma^\perp \cap C_{\ell u}(G)$ by the remarks before Proposition 2.2.20. Let $f, g \in J_\sigma^\perp \cap C_{\ell u}(G)$ and $a \in G$. Then we have

$$s \circ (f \underset{u}{\times} g)(a) = \left\langle R_a(f \underset{u}{\times} g), s \right\rangle$$
$$= \left\langle R_a P_\sigma(f \overline{u} g), s \right\rangle$$
$$= \left\langle P_\sigma(R_a(f \overline{u} g)), s \right\rangle$$
$$= \left\langle \overline{u}(a) P_\sigma((R_a f) \overline{u}(R_a g)), s \right\rangle$$
$$= \left\langle \overline{u}(a)(R_a f \underset{u}{\times} R_a g), s \right\rangle$$
$$= \overline{u}(a) \left\langle R_a f, s \right\rangle \left\langle R_a g, s \right\rangle.$$

Now let $s' = w^* - \lim_{\beta} \overline{u}\,\delta_{y_\beta}$. Then

$$\left\langle f \underset{u}{\times} g, s \circ s' \right\rangle = \left\langle s \circ (f \underset{u}{\times} g), s' \right\rangle$$

$$= \lim_{\beta} \left\langle s \circ (f \underset{u}{\times} g), \overline{u}\,\delta_{y_\beta} \right\rangle$$

$$= \lim_{\beta} \overline{u}\,(y_\beta)\big(s \circ (f \underset{u}{\times} g)(y_\beta)\big)$$

$$= \lim_{\beta} \overline{u}\,(y_\beta) \left\langle R_{y_\beta} f, s \right\rangle \overline{u}\,(y_\beta) \left\langle R_{y_\beta} g, s \right\rangle$$

$$= \lim_{\beta} \left\langle s \circ f, \overline{u}\,\delta_{y_\beta} \right\rangle \left\langle s \circ g, \overline{u}\,\delta_{y_\beta} \right\rangle$$

$$= \left\langle f, s \circ s' \right\rangle \left\langle g, s \circ s' \right\rangle$$

which gives $s \circ s' \in \Pi_\sigma$. It follows that Π_σ is a compact left topological semigroup and has an idempotent [23]. The structure stated in the theorem now follows from [71]. $\qquad\square$

Having shown that Π_σ is a semigroup above, one can, in principle, reduce the later arguments concerning the semigroup properties of Π_σ, whenever $u \in J_\sigma^\perp$ is a character, to the case in which σ is a probability measure in the following way. We can write $\sigma = u.\sigma_1$ where σ_1 is a probability measure on G, and define a linear isometry $T : J_\sigma^\perp \cap C_{\ell u}(G) \longrightarrow J_{\sigma_1}^\perp \cap C_{\ell u}(G)$ by

$$(Tf)(x) = u(x^{-1})f(x)$$

for $f \in J_\sigma^\perp \cap C_{\ell u}(G)$ and $x \in G$. Then the dual map T^* satisfies $T^*(\delta_x) = \overline{u}\delta_x$ for $x \in G$ and restricts to semigroup homeomorphisms $\tau : \mathcal{D}_{\sigma_1} \longrightarrow \mathcal{D}_\sigma$ and $\tau : \Pi_{\sigma_1} \longrightarrow \Pi_\sigma$ such that τ is equivariant:

$$\tau(x.s) = x.\tau(s) \qquad (x \in G, s \in \Pi_{\sigma_1})$$

where $x.s = R_{x^{-1}}^*(s)$ and $x.\tau(s) = \overline{\langle R_{x^{-1}}(u), \tau(s) \rangle} R_{x^{-1}}^*(\tau(s))$. Although it is simpler to deduce the semigroup results for Π_{σ_1}, we prefer a direct approach involving u to avoid lengthy conversion of the results for σ_1 to those for σ via τ.

Example 2.3.4. If G is a compact group, then $\mathcal{D}_\sigma = \{\overline{u}\delta_x : x \in G\}^- = \{\overline{u}\delta_x : x \in G\}$ since the map $x \in G \mapsto \overline{u}\delta_x \in (J_\sigma^\perp)^*$ is continuous. By Proposition 2.2.21, we have $\Pi_\sigma = \mathcal{D}_\sigma$ which is now a compact group and a homomorphic image of G.

Example 2.3.5. If $G = (\mathbb{Z}, +)$ and $\sigma = \delta_0$, then $J_\sigma^\perp = \ell^\infty(\mathbb{Z})$ and $\Pi_\sigma = \beta\mathbb{Z}$ is the Stone-Čech compactification of $\mathbb{Z}$. In fact, for any locally compact group G, we have $J_{\delta_e}^\perp = L^\infty(G)$ and Π_{δ_e} is the spectrum of $C_{\ell u}(G)$ which has been studied in [57].

Example 2.3.6. If $G = SL(2, \mathbb{R})$ and σ is a non-degenerate spread out probability measure on G, then $\Pi_\sigma = G/K$ is the circle S^1 where K is the compact subgroup of rotations (cf.[68, p. 211]). We recall that a measure σ is *spread out* if there exists $n \in \mathbb{N}$ such that σ^n and λ are not mutually singular.

2.4. Almost periodic harmonic functions

Having studied the uniformly continuous bounded σ-harmonic functions and their Poisson space Π_σ, we now consider the smaller class of almost periodic harmonic functions and their algebraic structures.

A function $f \in L^\infty(G)$ is called *almost periodic* (resp. *weakly almost periodic*) if the set $R_f = \{R_x f : x \in G\}$ of its right translates is relatively compact in the norm topology (resp. weak topology) of $L^\infty(G)$. The definition is equivalent to saying that the set of left translates $L_f = \{L_x f : x \in G\}$ is relatively compact. The almost periodic and weakly almost periodic functions form C^*-subalgebras of $L^\infty(G)$, and will be denoted by $\mathcal{AP}(G)$ and $\mathcal{WAP}(G)$ respectively. Let $C_u(G) = C_{\ell u}(G) \cap C_{ru}(G)$. It is well-known that $\mathcal{AP}(G) \subset \mathcal{WAP}(G) \subset C_u(G)$. See also [9]. If $u \in J_\sigma^\perp$ is a continuous character on G, then it also belongs to $\mathcal{AP}(G)$ in which case $J_\sigma^\perp \cap \mathcal{AP}(G)$ is a C^*-subalgebra of $J_\sigma^\perp$. We have the following more general result.

Proposition 2.4.1. *If the identity* $u \in J_\sigma^\perp$ *is in* $\mathcal{AP}(G)$, *then* $J_\sigma^\perp \cap \mathcal{AP}(G)$ *is a* C^*-*subalgebra of* $J_\sigma^\perp$. *The same result holds for* $\mathcal{WAP}(G)$.

Proof. Let $f, g \in J_\sigma^\perp \cap \mathcal{AP}(G)$. We need to show that $f \underset{u}{\times} g \in J_\sigma^\perp \cap \mathcal{AP}(G)$, that is, $R_{f \underset{u}{\times} g} = \{R_x(f \underset{u}{\times} g) : x \in G\}$ is relatively compact in $J_\sigma^\perp$. For $x \in G$, let $\widetilde{\tau}_x : J_\sigma^\perp \to J_\sigma^\perp$ be the algebra homomorphism as defined in (2.5):

$$\langle \widetilde{\tau}_x f, s \rangle = \overline{\langle R_x u, s \rangle} \langle R_x f, s \rangle$$

for $s \in \Pi_\sigma$. Since $u \in \mathcal{AP}(G)$, $\{R_x u : x \in G\}$ is relatively compact in $J_\sigma^\perp \cap C_{\ell u}(G) \simeq C(\Pi_\sigma)$ and it follows from the above identity and the Arzelà-Ascoli Theorem that $\{\widetilde{\tau}_x f : x \in G\}$ is relatively compact in $J_\sigma^\perp \cap C_{\ell u}(G)$.

Another application of the Arzelà-Ascoli Theorem to the identity

$$\langle \tilde{\tau}_x(f \underset{u}{\times} g), s \rangle = \langle \tilde{\tau}_x f, s \rangle \langle \tilde{\tau}_x g, s \rangle \quad (s \in \Pi_\sigma)$$

(see the proof of Lemma 2.2.10) implies that $\{\tilde{\tau}_x(f \underset{u}{\times} g) : x \in G\}$ is relatively compact in $J_\sigma^\perp \cap C_{\ell u}(G)$. Finally

$$\langle R_x(f \underset{u}{\times} g), s \rangle = \langle R_x u, s \rangle \langle \tilde{\tau}_x(f \underset{u}{\times} g), s \rangle \quad \text{for} \quad s \in \Pi_\sigma$$

again implies that $\{R_x(f \underset{u}{\times} g) : x \in G\}$ is relatively compact in $J_\sigma^\perp \cap C_{\ell u}(G)$. Hence $f \underset{u}{\times} g \in \mathcal{AP}(G)$.

For the case of $u \in \mathcal{WAP}(G)$, we note that for bounded sequences in $C(\Pi_\sigma)$, weak convergence is the same as pointwise convergence. Using the same identities as before, one can show that $\{R_x(f \underset{u}{\times} g) : x \in G\}$ is relatively weakly compact in $J_\sigma^\perp \cap C_{\ell u}(G)$ whenever $f, g \in J_\sigma^\perp \cap \mathcal{WAP}(G)$. □

Given $u \in J_\sigma^\perp \cap C_{\ell u}(G)$, we have shown in Theorem 2.2.17 that there is a net $\{\mu_\alpha\}$ in the convex hull co $\{\sigma^n : n \geq 1\}$ such that $(\mu_\alpha * f \overline{u} g)$ converges to $f \underset{u}{\times} g$ in the weak*-topology $\sigma(L^\infty(G), L^1(G))$, for every $f, g \in J_\sigma^\perp \cap C_{\ell u}(G)$.

Lemma 2.4.2. (i) *Given* $u \in \mathcal{WAP}(G)$ *and the net* $\{\mu_\alpha\}$ *in Theorem 2.2.17, then the net* $(\mu_\alpha * f \overline{u} g)$ *converges weakly to* $f \underset{u}{\times} g$ *for every* $f, g \in J_\sigma^\perp \cap \mathcal{WAP}(G)$.

(ii) *Given* $u \in \mathcal{AP}(G)$, *the net* $\mu_\alpha * f \overline{u} g$ *norm-converges to* $f \underset{u}{\times} g$ *for every* $f, g \in J_\sigma^\perp \cap \mathcal{AP}(G)$.

Proof. It suffices to show (i). Similar arguments apply to (ii). Let $h = f \overline{u} g$ and $K_h = \{\sum_{i=1}^n \lambda_i L_{x_i} h : x_i \in G, \sum_{i=1}^n |\lambda_i| \leq 1\}$. Then $\mu_\alpha * h \in \overline{K}_h$ as in Theorem 2.2.17, where the pointwise closure $\overline{K}_h$ is compact in the weak topology of $C_{\ell u}(G)$ since $h \in \mathcal{WAP}(G)$. Hence the weak topology agrees with the weak*-topology of $\overline{K}_h$ which completes the proof. □

Definition 2.4.3. Whenever $J_\sigma^\perp \cap \mathcal{WAP}(G)$ or $J_\sigma^\perp \cap \mathcal{AP}(G)$ is a C^*-subalgebra of $(J_\sigma^\perp, \underset{u}{\times})$, we denote their respective pure state spaces by Π_σ^w and Π_σ^a. In this case, the action of G on them is understood to be the natural one induced by the action $(x, t) \in G \times \Pi_\sigma \mapsto x \cdot t = \overline{\langle R_x u, t \rangle} R_{x^{-1}}^*(t)$.

Let X be a compact Hausdorff space with uniformity $\mathcal{U}$. An action of G on X is *equicontinuous* if for any $y \in X$ and $U \in \mathcal{U}$, there exists $V \in \mathcal{U}$ such that $(x, y) \in V$ implies $(a \cdot x, a \cdot y) \in U$ for all $a \in G$.

Let $u \in J_\sigma^\perp \cap C_{\ell u}(G) \simeq C(\Pi_\sigma)$. We recall that the weak* topology on $\left(J_\sigma^\perp \cap C_{\ell u}(G)\right)^*$ is defined by the seminorms $n_f : s \mapsto |\langle f, s\rangle|$ where $f \in J_\sigma^\perp \cap C_{\ell u}(G)$ and $s \in \left(J_\sigma^\perp \cap C_{\ell u}(G)\right)^*$. Since $\langle \tilde{\tau}_x f, s\rangle = \langle f, x^{-1} \cdot s\rangle$ for $f \in J_\sigma^\perp \cap C_{\ell u}(G)$ and $s \in \Pi_\sigma$, by (2.5), we see that the action of G on Π_σ (always equipped with the weak*-topology) is equicontinuous if, and only if, $\{\tilde{\tau}_x f : x \in G\}$ is equicontinuous in $C(\Pi_\sigma)$ for each $f \in J_\sigma^\perp \cap C_{\ell u}(G)$. This leads to the following result.

Proposition 2.4.4. *Let* $u \in J_\sigma^\perp \cap C_{\ell u}(G)$. *If the action of* G *on* Π_σ *is equicontinuous, then* $J_\sigma^\perp \cap C_{\ell u}(G) = J_\sigma^\perp \cap \mathcal{AP}(G)$. *If* $u \in \mathcal{AP}(G)$, *then the action of* G *on* Π_σ^a *is equicontinuous.*

Proof. Let $f \in J_\sigma^\perp \cap C_{\ell u}(G)$. Then the equicontinuity of the action implies that $\{\tilde{\tau}_x f : x \in G\}$ is equicontinuous in $C(\Pi_\sigma)$, and hence relatively compact since it is bounded. So $f \in \mathcal{AP}(G)$ by arguments similar to the proof of Proposition 2.4.1.

Given $u \in \mathcal{AP}(G)$, then for each $f \in J_\sigma^\perp \cap \mathcal{AP}(G) \simeq C(\Pi_\sigma^a)$, the set $\{\tilde{\tau}_x f : x \in G\}$ is relatively norm-compact in $C(\Pi_\sigma^a)$, and in particular, equicontinuous. Therefore G acts on Π_σ^a equicontinuously as before. $\qquad\square$

Let S_σ be the state space of $J_\sigma^\perp \cap C_{\ell u}(G)$ and let G act on S_σ via $(x, t) \mapsto x \cdot t = \overline{\langle R_{x^{-1}} u, t\rangle} R_{x^{-1}}^*(t)$. Given $u \in \mathcal{WAP}(G)$, we let S_σ^w be the state space of the C^*-algebra $J_\sigma^\perp \cap \mathcal{WAP}(G)$. The above action of G induces a natural action on S_σ^w.

Proposition 2.4.5. (i) *There is a locally convex topology* $\mathcal{T}$ *for the dual pair* $\left(J_\sigma^\perp \cap \mathcal{WAP}(G)\right)^*$ *and* $J_\sigma^\perp \cap \mathcal{WAP}(G)$ *such that the action of* G *on* $(S_\sigma^w, \mathcal{T})$ *is equicontinuous.*

(ii) *If there is a locally convex topology* $\mathcal{T}$ *on the dual pair* $\left(J_\sigma \cap C_{\ell u}(G)\right)^*$ *and* $J_\sigma^\perp \cap C_{\ell u}(G)$ *such that the action of* G *on* $(S_\sigma, \mathcal{T})$ *is equicontinuous, then* $J_\sigma^\perp \cap \mathcal{WAP}(G) = J_\sigma^\perp \cap C_{\ell u}(G)$.

Proof. (i) For $f \in J_\sigma^\perp \cap \mathcal{WAP}(G)$, define a seminorm p_f on $\left(J_\sigma^\perp \cap \mathcal{WAP}(G)\right)^*$ by

$$p_f(t) = \sup\{|\langle \tilde{\tau}_x f, t\rangle| : x \in G\}.$$

Let $\mathcal{T}$ be the locally convex topology on $\left(J_\sigma^\perp \cap \mathcal{WAP}(G)\right)^*$ defined by the seminorms $\{p_f : f \in J_\sigma^\perp \cap \mathcal{WAP}(G)\}$. Since $f \in \mathcal{WAP}(G)$ implies that $\{\tilde{\tau}_x f :$

$x \in G$} is relatively compact in $J_\sigma^\perp \cap WAP(G)$, it follows from Mackey-Arens Theorem [69] that $\mathcal{T}$ is a topology for the dual pair $((J_\sigma^\perp \cap WAP(G))^*, (J_\sigma \cap WAP(G)))$ and G acts equicontinuously on $(S_\sigma^w, \mathcal{T})$.

(ii) As in (2.4), for each $x \in G$, there is a map $\tau_x : S_\sigma \to (J_\sigma^\perp \cap C_{\ell u}(G))^*$ such that

$$\langle R_x u, s \rangle \langle f, \tau_x(s) \rangle = \langle R_x f, s \rangle$$

for $f \in J_\sigma^\perp \cap C_{\ell u}(G)$ and $s \in S_\sigma$. Write $E = (J_\sigma^\perp \cap C_{\ell u}(G))^*$ and let $\mathcal{G} = \{\tau_x : x \in G\} \subset E^{S_\sigma}$. Then the convex hull co $\mathcal{G}$ has the same closure $\overline{\text{co}}\,\mathcal{G}$ in $(E, w^*)^{S_\sigma}$ and in $(E, \mathcal{T})^{S_\sigma}$. Since the action of G on $(S_\sigma, \mathcal{T})$ is equicontinuous, the family $\mathcal{G}$ is equicontinuous in $(E, \mathcal{T})^{(S_\sigma, \mathcal{T})}$. Therefore $\overline{\text{co}}\,\mathcal{G}$ consists of continuous maps from $(S_\sigma, \mathcal{T})$ to $(E, \mathcal{T})$ which are also continuous maps from (S_σ, w^*) to (E, w^*), by convexity of S_σ. Hence the closure $\overline{\mathcal{G}}$ in $(E, w^*)^{S_\sigma}$ consists of maps continuous from (S_σ, w^*) to (E, w^*). Let $f \in J_\sigma^\perp \cap C_{\ell u}(G) \approx C(\Pi_\sigma)$. Then the map $\gamma \in \overline{\mathcal{G}} \mapsto f_\gamma \in C(\Pi_\sigma)$, where $f_\gamma(s) = \langle f, \gamma(s) \rangle$ for $s \in \Pi_\sigma$, is continuous in the weak topology of $C(\Pi_\sigma)$. Therefore $\{f_\gamma : \gamma \in \overline{\mathcal{G}}\}$ is weakly compact in $C(\Pi_\sigma)$ and it follows from $\{\widetilde{\tau}_x f : x \in G\} \subset \{f_\gamma : \gamma \in \overline{\mathcal{G}}\}$ that $f \in WAP(G)$. $\qquad\square$

We do not know if S_σ can be replaced by Π_σ in Proposition 2.4.5 (ii).

Proposition 2.4.6. *Given that $u \in WAP(G)$ (resp. $AP(G)$), then there is a G-invariant probability measure on Π_σ^w (resp. Π_σ^a).*

Proof. Let $\mathcal{T}$ be the locally convex topology on $(J_\sigma^\perp \cap WAP(G))^*$ defined by the seminorms $\{p_f : f \in J_\sigma^\perp \cap WAP(G)\}$, as in the proof of Proposition 2.4.5. We observe that the action of G on $(S_\sigma^w, \mathcal{T})$ is distal, that is, given a net (a_α) in G with $\lim_\alpha (a_\alpha \cdot s - a_\alpha \cdot t) = 0$ in the topology $\mathcal{T}$, we have $s = t$. Indeed, for any $\varepsilon > 0$ and $f \in J_\sigma^\perp \cap WAP(G)$, there exists α_0 such that $p_f(a_{\alpha_0} \cdot s - a_{\alpha_0} \cdot t) < \varepsilon$, in other words, $\sup\{|\langle \widetilde{\tau}_x f, a_{\alpha_0} \cdot s - a_{\alpha_0} \cdot t \rangle| : x \in G\} < \varepsilon$ and for $x = a_{\alpha_0} \in G$, we have $|\langle f, s - t \rangle| < \varepsilon$ which gives $s = t$. By Ryll-Nardzewski fixed point theorem, there exists $\varphi \in (J_\sigma^\perp \cap WAP(G))^*$ such that $\varphi(u) = 1 = \|\varphi\|$ and $a \cdot \varphi = \varphi$ for all $a \in G$. So the probability measure on Π_σ^w representing φ is the required G-invariant measure. The case for Π_σ^a is proved similarly. $\qquad\square$

We have studied conditions in which the Poisson space Π_σ coincides with Π_σ^a or Π_σ^w. We now study their relationships further under the condition that $J_\sigma^\perp$ is translation invariant.

Definition 2.4.7. We call a semitopological semigroup S *amenable* if $C_b(S)$ has an *invariant mean* μ, that is, there exists a positive $\mu \in C_b(S)^*$ such that $\|\mu\| = 1$ and $\mu({}_sf) = \mu(f_s) = \mu(f)$ for $f \in C_b(S)$ and $s \in S$, where ${}_sf(t) = f(st)$ and $f_s(t) = f(ts)$ for $t \in S$.

Given $g \in J_\sigma^\perp \cap \mathcal{WAP}(G) \subset J_\sigma^\perp \cap C_{\ell u}(G)$ and $m \in \left(J_\sigma^\perp \cap \mathcal{WAP}(G)\right)^*$, we can define, as in (2.2), $m \circ f \in J_\sigma^\perp \cap \mathcal{WAP}(G)$ by

$$m \circ f(x) = \langle R_x f, m \rangle$$

for $x \in G$. Then $\left(J_\sigma^\perp \cap \mathcal{WAP}(G)\right)^*$ is a Banach algebra in the following product

$$\langle f, m \circ n \rangle = \langle m \circ f, n \rangle$$

where $f \in J_\sigma^\perp \cap \mathcal{WAP}(G)$ and $m, n \in \left(J_\sigma^\perp \cap \mathcal{WAP}(G)\right)^*$. Let $\mathcal{D}_\sigma^w$ be the restriction to $J_\sigma^\perp \cap \mathcal{WAP}(G)$ of the functionals in $\mathcal{D}_\sigma = \{\bar{u}\delta_x : x \in G\}^-$ (cf. Theorem 2.3.3).

Theorem 2.4.8. *Let $J_\sigma^\perp$ be translation invariant with identity u which is a continuous character on G. Then*

  (i) *$(\mathcal{D}_\sigma^w, \circ)$ is a compact semitopological semigroup where the product $\circ$ is that of $\left(J_\sigma^\perp \cap \mathcal{WAP}(G)\right)^*$ defined above;*

  (ii) *$(\Pi_\sigma^w, \circ)$ is an amenable closed subsemigroup of $\mathcal{D}_\sigma^w$.*

  (iii) *If $(\Pi_\sigma, \circ)$ is a compact semitopological semigroup, then $J_\sigma^\perp \cap \mathcal{WAP}(G) = J_\sigma^\perp \cap C_{\ell u}(G)$.*

Proof. (i) Arguments similar to those in the proof of Theorem 2.3.3 show that $(\mathcal{D}_\sigma^w, \circ)$ is a semigroup. We show that multiplication in $\mathcal{D}_\sigma^w$ is separately continuous. Let $f \in J_\sigma^\perp \cap \mathcal{WAP}(G)$ and write $uL_f = \{\bar{u}(x)L_x f : x \in G\}$. Let $\gamma \in \mathcal{D}_\sigma^w$ with $\gamma = w^* - \lim_\alpha \bar{u}\delta_{x_\alpha}$. Then $\gamma \circ f(x) = \langle R_x f, \gamma \rangle = \lim_\alpha \langle R_x f, \bar{u}(x)\delta_{x_\alpha} \rangle = \lim_\alpha \bar{u}(x)L_{x_\alpha} f(x)$ for $x \in G$. So $\gamma \circ f$ is in the pointwise closure of uL_f. But uL_f is relatively weakly compact as $f \in \mathcal{WAP}(G)$, the pointwise and weak topology agree on the weak-closed convex hull of uL_f. Now let (γ_α) be a net in $\mathcal{D}_\sigma^w$ w^*-converging to $\gamma \in \mathcal{D}_\sigma^w$. Then $(\gamma_\alpha \circ f)$ converges to $\gamma \circ f$ pointwise in uL_f, and hence weakly. Therefore, for any $\gamma' \in \mathcal{D}_\sigma^w$, we have

$$\langle f, \gamma_\alpha \circ \gamma' \rangle = \langle \gamma_\alpha \circ f, \gamma' \rangle \to \langle \gamma \circ f, \gamma' \rangle = \langle f, \gamma \circ \gamma' \rangle$$

which gives separate w^*-continuity of $\circ$ on $\mathcal{D}_\sigma^w$.

  (ii) We show that Π_σ^w is amenable. By the proof of Proposition 2.4.6, there exists $\varphi \in C(\Pi_\sigma^w)^*$ such that $\varphi(u) = 1 = \|\varphi\|$ and $a \cdot \varphi = \varphi$ for all $a \in G$. Hence for $f \in J_\sigma^\perp \cap \mathcal{WAP}(G) \simeq C(\Pi_\sigma^w)$ and $a \in G$, we have

$$\langle f, \varphi \rangle = \langle f, a \cdot \varphi \rangle = \overline{\langle R_a u, \varphi \rangle} \langle R_a f, \varphi \rangle = \bar{u}(a) \langle R_a f, \varphi \rangle$$

by multiplicativity of u. Let $t \in \mathcal{D}_\sigma^w$. By separate continuity of $\circ$, we can define $f_t \in C(\Pi_\sigma^w)$ by $f_t(s) = \langle f, s \circ t \rangle$ for $s \in \Pi_\sigma^w$. Then for $t = \overline{u}\delta_a$, we have

$$f_t(s) = \overline{u}(a) \langle f, s \circ \delta_a \rangle = \overline{u}(a) \langle s \circ f, \delta_a \rangle = \overline{u}(a)(s \circ f)(a) = \overline{u}(a) \langle R_a f, s \rangle.$$

Hence $\langle f_t, \varphi \rangle = \overline{u}(a) \langle R_a f, \varphi \rangle = \langle f, \varphi \rangle$.

We note that the set $\{f_t : t \in \mathcal{D}_\sigma^w\}$ is weakly compact in $C(\Pi_\sigma^w)$ since it is bounded and pointwise compact, by compactness of $\mathcal{D}_\sigma^w$ and continuity of the map $t \in \mathcal{D}_\sigma^w \mapsto f_t \in C(\Pi_\sigma^w)$. It follows that, given $s \in \Pi_\sigma^w$ with $s = w^* - \lim_\alpha \overline{u}\delta_{a_\alpha}$, then $f_{\overline{u}\delta_{a_\alpha}} \to f_s$ weakly and $\langle f_s, \varphi \rangle = \lim_\alpha \langle f_{\overline{u}\delta_{a_\alpha}}, \varphi \rangle = \langle f, \varphi \rangle$.

Analogous to the proof of Proposition 2.4.6, we can find $\psi \in C(\Pi_\sigma^w)^*$ such that $\psi(u) = \|\psi\| = 1$, and

$$\langle f, \psi \rangle = \overline{u}(a) \langle L_a f, \psi \rangle$$

for $f \in J_\sigma^\perp \cap \mathcal{WAP}(G)$ and $a \in G$. As above, for $t \in \mathcal{D}_\sigma^w$, we define $_t f \in C(\Pi_\sigma^w)$ by $_t f(s) = \langle f, t \circ s \rangle$ for $s \in \Pi_\sigma^w$ and obtain $\langle _t f, \psi \rangle = \langle f, \psi \rangle$ by similar arguments.

Now we define $\mu \in C(\Pi_\sigma^w)^*$ by

$$\mu(f) = \langle \psi \cdot f, \varphi \rangle$$

where $\psi \cdot f \in C(\Pi_\sigma^w)$ is defined by

$$(\psi \cdot f)(s) = \langle f_s, \psi \rangle$$

for $s \in \Pi_\sigma^w$. Then it is straightforward to verify that μ is an invariant mean.

(iii) Suppose multiplication in Π_σ is separately continuous. We show that $J_\sigma^\perp \cap C_{\ell u}(G) = J_\sigma^\perp \cap \mathcal{WAP}(G)$. Let $f \in J_\sigma^\perp \cap C_{\ell u}(G) \approx C(\Pi_\sigma)$. We show that $\{\overline{u}(a) R_a f : a \in G\}$ is relatively weakly compact which would imply $f \in \mathcal{WAP}(G)$. As in (ii), separate continuity implies that the set $\{f_t : t \in \Pi_\sigma\}$ is pointwise compact in $C(\Pi_\sigma)$ and therefore the pointwise closed convex hull $K = \overline{co} \{f_t : t \in \Pi_\sigma\}$ is weakly compact. To complete the proof, it suffices to show that $\overline{u}(a) R_a f \in K$ for all $a \in G$. By Proposition 2.2.20, $\overline{u}\delta_a = w^* - \lim_\alpha \mu_\alpha$ where $\mu_\alpha = \sum_j \lambda_j^\alpha s_j^\alpha$ is a convex combination of the pure states

$s_j^\alpha \in \Pi_\sigma$. It follows that, for $s \in \Pi_\sigma$,

$$\langle \overline{u}(a)R_a f, s \rangle = \langle s \circ f, \overline{u}\delta_a \rangle$$

$$= \lim_\alpha \langle s \circ f, \sum_j \lambda_j^\alpha s_j^\alpha \rangle$$

$$= \lim_\alpha \langle \sum_j \lambda_j^\alpha f_{s_j^\alpha}, s \rangle$$

which gives $\overline{u}(a)R_a f \in K$. $\qquad \square$

As in the case of $J_\sigma^\perp \cap \mathcal{WAP}(G)$, we can define the multiplication $\circ$ on $\left(J_\sigma^\perp \cap AP(G)\right)^*$ which then becomes a Banach algebra.

Theorem 2.4.9. *Let $J_\sigma^\perp$ be translation invariant with identity u which is a continuous character on G. Then $(\Pi_\sigma^a, \circ)$ is a compact group and there is a continuous homomorphism from G onto a dense subgroup of Π_σ^a.*

Proof. As before, let $\mathcal{D}_\sigma^a$ be the restriction of $\{\overline{u}\delta_x : x \in G\}^-$ to $J_\sigma^\perp \cap AP(G)$. Then $(\mathcal{D}_\sigma^a, \circ)$ is a compact semitopological semigroup. We show $\mathcal{D}_\sigma^a$ is a group. Since $f \in J_\sigma^\perp \cap AP(G)$ implies that $uL_f = \{\overline{u}(x)L_x f : x \in G\}$ is relatively norm-compact in $J_\sigma^\perp \cap AP(G) \simeq C(\Pi_\sigma^a)$, by arguments similar to those in the proof of Theorem 2.4.8(i) with the fact that $s_\alpha \to s$ in Π_σ^a implies $s_\alpha \circ f \to s \circ f$ in norm, one can show that the multiplication $\circ$ is jointly continuous in $\mathcal{D}_\sigma^a$. As $(\{\overline{u}\delta_x : x \in G\}, \circ)$ is a group dense in $\mathcal{D}_\sigma^a$, it follows that $(\mathcal{D}_\sigma^a, \circ)$ is a group [40, p. 13].

We note that the restriction map $\Pi_\sigma^w \to \Pi_\sigma^a$ is a continuous homomorphism and since Π_σ^w is amenable by Theorem 2.4.8, Π_σ^a is also amenable. By [8, p. 87], the minimal two-sided ideal K in Π_σ^a must be a compact group. Let θ be the identity of K. Then $\theta = \theta \circ \theta = \theta \circ \widetilde{\theta}$ where $\widetilde{\theta}$ is the identity of $\mathcal{D}_\sigma^a$. By cancellation, $\widetilde{\theta} = \theta$ and hence $\Pi_\sigma^a = K$ is a group with identity θ.

Finally we observe that, given $s \in \Pi_\sigma^a$, we have $s \circ \overline{u}\delta_s = \overline{u}(x)R_x^*(s) = \overline{\langle R_x u, s \rangle} R_x^*(s) \in \Pi_\sigma^a$ by multiplicativity of u and Lemma 2.2.10. Therefore we can define a continuous map $h : G \to \Pi_\sigma^a$ by $h(x) = \theta \circ \overline{u}\delta_x$ for $x \in G$. Then h is a homomorphism since $h(xy) = \overline{u}(xy)\delta_{xy} = (\theta \circ \overline{u}\delta_x) \circ \overline{u}\delta_y = (\theta \circ \overline{u}\delta_x) \circ (\theta \circ \overline{u}\delta_y)$. Since $\Pi_\sigma^a \subset \{\overline{u}\delta_x : x \in G\}^-$, we have $h(G)$ dense in Π_σ^a. $\qquad \square$

We conclude this section by showing that Π_σ^a is (isomorphic to) the minimal two-sided ideal of Π_σ^w.

Theorem 2.4.10. *Let $J_\sigma^\perp$ be translation invariant with identity u which is a continuous character on G. Then Π_σ^a is topologically isomorphic to the minimal two-sided ideal M of Π_σ^w.*

Proof. The restriction map $r : \Pi^w_\sigma \to \Pi^a_\sigma$ is easily seen to be a continuous homomorphism. We show that r restricts to the required isomorphism on M.

By [8, p. 87], M is a compact group with identity θ say. We note that $M \circ \overline{u}\delta_x \subset M$ for every $x \in G$. Indeed, we have $\theta \circ \overline{u}\delta_x \in \Pi^w_\sigma$ as before and $s \in M$ implies

$$s \circ \overline{u}\delta_x = (s \circ \theta) \circ \overline{u}\delta_x = s \circ (\theta \circ \overline{u}\delta_x) \in M$$

since M is an ideal in Π^w_σ. Therefore, as in Theorem 2.4.8, we can define a continuous homomorphism $h : G \to M$ by $h(x) = \theta \circ \overline{u}\delta_s$ with $h(G)$ dense in M.

The map h induces a linear isometry $f \in C(M) \mapsto f \cdot h \in C_b(G)$ with range in $J^\perp_\sigma \cap AP(G)$ where we define $(f \cdot h)(x) = \overline{u}(x^{-1}f(h(x)))$ for $x \in G$. To see the latter, we first show that $f \cdot h \in AP(G)$. For $t \in M$, define $f_t \in C(M)$ by $f_t(s) = f(s \circ t)$ for $s \in M$. Then for $x \in G$, we have

$$R_s(f \cdot h) = \overline{u}(x^{-1})f_{h(x)} \cdot h.$$

Since M is compact, the set $\{f_t \cdot h : t \in M\}$ is norm-compact in $C_b(G)$ and so is its subset $\{f_{h(x)} \cdot h : x \in G\}^-$. It follows that $\{R_x(f \cdot h) : x \in G\}$ is relatively norm-compact in $C_b(G)$ and $f \cdot h \in AP(G)$. To see that $f \cdot h \in J^\perp_\sigma$, we let $\widetilde{f} \in C(\Pi^w_\sigma) \simeq J^\perp_\sigma \cap WAP(G)$ be an extension of $f \in C(M)$. Then for $x \in G$, we have

$$(\sigma * f \cdot h)(x) = \int_G (f \cdot h)(y^{-1}x) d\sigma(y)$$

$$= \int_G \overline{u}(x^{-1}y)f\big(h(y^{-1}x)\big) d\sigma(y)$$

$$= \int_G \overline{u}(x^{-1}y) \langle f, \theta \circ \overline{u}\delta_{y^{-1}x} \rangle d\sigma(y)$$

$$= \int_G \overline{u}(x^{-1}y)\overline{u}(y^{-1}x) \langle \widetilde{f}, \theta \circ \delta_{y^{-1}x} \rangle d\sigma(y)$$

$$= \int_G (\theta \circ \widetilde{f})(y^{-1}x) d\sigma(y)$$

$$= \theta \circ \widetilde{f}(x) = \overline{u}(x^{-1}) \langle f, \theta \circ \overline{u}\delta_x \rangle = (f \cdot h)(x)$$

since $\theta \circ \widetilde{f} \in J^\perp_\sigma$. So $f \cdot h \in J^\perp_\sigma$.

Now we show that the restriction map r is injective on M. Let $s, t \in M$ with $s \neq t$. Then there exists $f \in C(M)$ with $f(s) \neq f(t)$. Note that $f \cdot h = \theta \circ f$ as $(f \cdot h)(x) = \overline{u}(x^{-1}) f(\theta \circ \overline{u}\delta_x) = (\theta \circ f)(x)$ for $x \in G$. Since $f \cdot h \in J_\sigma^\perp \cap \mathcal{AP}(G) \simeq C(\Pi_\sigma^a)$, we have $\langle f \cdot h, r(s) \rangle = \langle f \cdot h, s \rangle = \langle \theta \circ f, s \rangle = \langle f, \theta \circ s \rangle = \langle f, s \rangle \neq \langle f, t \rangle = \langle f \cdot h, r(t) \rangle$ which gives $r(s) \neq r(t)$.

Finally, for $x \in G$, we have $r(\theta \circ \overline{u}\delta_x) = r(\theta) \circ \overline{u}\delta_x$ and $r(\theta)$ is the identity of Π_σ^a. So $r(M)$ contains the dense subset $\{r(\theta) \circ \overline{u}\delta_x : x \in G\}$ of Π_σ^a which implies that r is onto Π_σ^a. $\qquad\square$

2.5. Distal harmonic functions

In this section, we study harmonic functions which are distal. Let $f \in C_{\ell u}(G)$ and let $\overline{R}_f$ denote the pointwise closure of $R_f = \{R_x f : x \in G\}$. We call f a *right distal function* if the action $(x, h) \in G \times \overline{R}_f \mapsto R_x h \in \overline{R}_f$ is *right distal* which means that, if $h_1, h_2 \in \overline{R}_f$ and

$$\lim_\alpha R_{x_\alpha} h_1 = \lim_\alpha R_{x_\alpha} h_2 \quad \text{(pointwise)}$$

then $h_1 = h_2$. We note that $\mathcal{AP}(G) \subset D(G)$, but the reverse inclusion is false, for instance, the function $f(n) = \exp(2\pi i n^2 \theta)$ is distal on $\mathbb{Z}$, but not almost periodic if θ is irrational (cf. [50]). The set $D(G)$ of all right distal functions on G forms a translation invariant C^*-subalgebra of $C_{\ell u}(G)$. If $J_\sigma^\perp$ is left-translation invariant, then so is $J_\sigma^\perp \cap D(G)$.

As before, we assume that the identity $u \in J_\sigma^\perp$ is a continuous character. Then $u \in J_\sigma^\perp \cap D(G)$. Let $S = \{\mu \in (J_\sigma^\perp \cap D(G))^* : \|\mu\| = \mu(u) = 1\}$ be the state space of $J_\sigma^\perp \cap D(G)$ and let

$$\rho : (J_\sigma^\perp \cap C_{\ell u}(G))^* \to (J_\sigma^\perp \cap D(G))^*$$

be the restriction map. Let $\Pi_\sigma^d = \rho(\Pi_\sigma)$ which is w^*-compact.

Since ρ is onto by the Hahn-Banach Theorem, we see that S is the w^*-closed convex hull of Π_σ^d and hence Π_σ^d contains the extreme points of S.

Lemma 2.5.1. *Let* $s \in \Pi_\sigma^d$ *and* $f \in J_\sigma^\perp \cap D(G)$. *Then the function* $(s \circ f)(x) = \langle R_x f, s \rangle$ *is in* $J_\sigma^\perp \cap D(G)$.

Proof. We have $s \circ f \in J_\sigma^\perp$ as before. We show that $s \circ f \in D(G)$. Let $\widetilde{s} \in \Pi_\sigma$ be such that $\rho(\widetilde{s}) = s$. By Proposition 2.2.20, $\widetilde{s}$ is the w^*-limit of a net $(\overline{u}\delta_{x_\alpha})_\alpha$ with $x_\alpha \in G$. We note that $\widetilde{s}$ admits an extension to

$\widetilde{\widetilde{s}} \in C_{\ell u}(G)^*$ such that $\widetilde{\widetilde{s}}(1) = \lim_{\beta} \overline{u}(x_\beta)$ for a subnet (x_β) of (x_α), and that (δ_{x_β}) w^*-converges to some $\theta \in \left(J_\sigma^\perp \cap C_{\ell u}(G)\right)^*$. Then $\widetilde{s} = \widetilde{\widetilde{s}}(1)\theta$ and $s \circ f = \widetilde{s} \circ f = \widetilde{\widetilde{s}}(1)(\theta \circ f) \in D(G)$ by [8, Theorem 4.6.3]. □

It follows from Lemma 2.5.1 that $(\Pi_\sigma^d, \circ)$ is a compact left topological semigroup where the product $\circ$ is defined, as before, by

$$\langle f, s \circ t \rangle = \langle s \circ f, t \rangle$$

for $f \in J_\sigma^\perp \cap D(G)$ and $s, t \in \Pi_\sigma^d$.

Theorem 2.5.2. Let $J_\sigma^\perp$ be translation invariant with identity u which is a continuous character on G. Then Π_σ^d is a compact left topological semigroup such that

$$s \circ \varepsilon \circ \theta = s \circ \theta$$

for $s, \theta \in \Pi_\sigma^d$ and for any idempotent ε in Π_σ^d.

  (i) If Π_σ^d has a right identity, then Π_σ^d is isomorphic to a direct product of a left zero semigroup and a group.
  (ii) If Π_σ^d has a left identity, then Π_σ^d is isomorphic to a direct product of a right zero semigroup and a group.
  (iii) If Π_σ^d has an identity, then it is a group.

Proof. Let $\widetilde{s}, \widetilde{\theta} \in \Pi_\sigma$ with $\rho(\widetilde{s}) = s$ and $\rho(\widetilde{\theta}) = \theta$. Since $\{t \in \Pi_\sigma : \rho(t) = \varepsilon\}$ is a closed subsemigroup of Π_σ, it contains an idempotent $\widetilde{\varepsilon}$ such that $\rho(\widetilde{\varepsilon}) = \varepsilon$. Let $f \in J_\sigma^\perp \cap D(G)$. Since $\{\overline{u}\delta_x : x \in G\}$ is weak*-dense in Π_σ by Proposition 2.2.20, $(\widetilde{s} \circ \widetilde{\varepsilon}) \circ f$ and $\widetilde{s} \circ f$ are in the pointwise closure of $\{\overline{u}(x)L_{x^{-1}}f : x \in G\}$. Since $(\widetilde{s} \circ \widetilde{\varepsilon}) \circ \widetilde{\varepsilon} \circ f = (\widetilde{s} \circ \widetilde{\varepsilon}) \circ f$, we have $(\widetilde{s} \circ \widetilde{\varepsilon}) \circ f = \widetilde{s} \circ f$ because f is distal. Consequently, $\langle f, \widetilde{s} \circ \widetilde{\varepsilon} \circ \overline{u}\delta_x \rangle = \langle f, \widetilde{s} \circ \overline{u}\delta_x \rangle$ for all $x \in G$. Passing to weak*-limit, we have $\langle f, \widetilde{s} \circ \widetilde{\varepsilon} \circ \widetilde{\theta} \rangle = \langle f, \widetilde{s} \circ \widetilde{\theta} \rangle$. As $f \in J_\sigma^\perp \cap D(G)$ was arbitrary, we have $s \circ \varepsilon \circ \theta = s \circ \theta$.
(i) If Π_σ^d has a right identity, then by above, we have $\varepsilon \circ \theta = \theta$ for every θ in Π_σ^d and every idempotent ε in Π_σ^d. By Ruppert's result [71], Π_σ^d cannot have any proper left ideal and is therefore isomorphic to a direct product of a left zero semigroup and a group (cf. [8, 1.2.19]).   (ii) is proved similarly and (iii) follows from (i) and (ii). □

Remark. If σ is a probability measure in Theorem 2.5.2 and $J_\sigma^\perp$ a C^*-subalgebra of $L^\infty(G)$, then Π_σ^d is a group since δ_e is an identity of Π_σ^d in this case.

Given $s \in \Pi_\sigma$ with $s = w^* - \lim_\alpha \overline{u}(x_\alpha)\delta_{x_\alpha}$ as in Proposition 2.2.20, we noted in the proof of Lemma 2.5.1 that s admits an extension to $\widetilde{s} \in C_{\ell u}(G)^*$ such that $\widetilde{s}(1) = \lim_\beta \overline{u}(x_\beta)$ for a subnet (x_β) of (x_α). We write $\overline{s}(1) = \overline{\widetilde{s}(1)} = \lim_\beta u(x_\beta)$.

Let $f \in J_\sigma^\perp \cap C_{\ell u}(G)$ and let $\overline{L}_f$ be the pointwise closure of $L_f = \{L_x f : x \in G\}$. Suppose $J_\sigma^\perp$ is also left-translation invariant. Then $L_f \subset J_\sigma^\perp$ and the set

$$L_f^\sigma = \{\overline{s}(1)(s \circ f) : s \in \Pi_\sigma\}$$

is contained in $\overline{L}_f$ since for $s = w^* - \lim_\alpha \overline{u}(x_\alpha)\delta_{x_\alpha}$ and $x \in G$, we have

$$\overline{s}(1)(s \circ f)(x) = \lim_\beta u(x_\beta) \langle R_x f, \overline{u}(x_\beta)\delta_{x_\beta} \rangle$$

$$= \lim_\beta \left(L_{x_\beta^{-1}} f \right)(x).$$

Note that for $a \in G$, we have $L_{a^{-1}}\big(\overline{s}(1)(s \circ f)\big) = \overline{t}(1)(t \circ f)$ where $t = \overline{\langle R_a u, s \rangle} R_a^* s \in \Pi_\sigma$.

Given $f \in J_\sigma^\perp \cap C_{\ell u}(G)$, we call f *left σ-distal* if the action $(x, h) \in G \times L_f^\sigma \mapsto L_{x^{-1}}h \in L_f^\sigma$ is *left distal* which means that, for $h_1, h_2 \in L_f^\sigma$,

$$\lim_\alpha L_{x_\alpha} h_1 = \lim_\alpha L_{x_\alpha} h_2 \quad \text{(pointwise)}$$

implies $h_1 = h_2$.

Lemma 2.5.3. *Let $J_\sigma^\perp$ be translation invariant with identity u which is a continuous character. Let G act transitively on Π_σ. Then Π_σ is isomorphic to a direct product of a right zero semigroup and a group.*

Proof. Let R be a minimal right ideal of Π_σ and let $p \in \Pi_\sigma$ be an idempotent such that $R = p\Pi_\sigma$. For $x \in G$, we have

$$x \cdot p = \overline{\langle R_{x^{-1}}u, p \rangle} \left(p \circ \delta_{x^{-1}}\right)$$

and hence $x \cdot p = x \cdot (p \circ p) = \overline{\langle R_{x^{-1}}u, p \rangle} \left(p \circ (p \circ \delta_{x^{-1}})\right) = \overline{\langle R_{x^{-1}}u, p \rangle} \langle R_{x^{-1}}u, p \rangle \left(p \circ (x \cdot p)\right) = p \circ (x \cdot p) \in R$. Given $s \in \Pi_\sigma$, we have $s = x \cdot p \in R$ for some $x \in G$ by transitivity of the group action. So $R = \Pi_\sigma$ and Π_σ has no proper right ideal. By Theorem 2.3.3, Π_σ has an idempotent and it follows from

[8, Theorem 1.2.19] that Π_σ is isomorphic to a direct product of a right zero semigroup and a group.   □

Theorem 2.5.4. *Let $J_\sigma^\perp$ be translation invariant with identity u which is a continuous character. If G acts transitively on Π_σ, then every $f \in J_\sigma^\perp \cap C_{\ell u}(G)$ is left σ-distal.*

Proof. Let $f \in J_\sigma^\perp \cap C_{\ell u}(G)$. By Lemma 2.5.3, we have $\Pi_\sigma \simeq E \times H$ where E is a right zero semigroup and H is a group. Hence for $s, t \in \Pi_\sigma$ and for any idempotent $p \in \Pi_\sigma$, we have

$$s \circ p \circ t = s \circ t.$$

Let $h_1, h_2 \in L_f^\sigma$ with $\lim_\alpha L_{x_\alpha^{-1}} h_1 = \lim_\alpha L_{x_\alpha^{-1}} h_2$ pointwise. We need to show $h_1 = h_2$. Let $h_1 = \bar{s}_1(1)(s_1 \circ f)$ and $h_2 = \bar{s}_2(1)(s_2 \circ f)$ for some $s_1, s_2 \in \Pi_\sigma$. Passing to a subnet, we may assume that the net $(\bar{u}\,\delta_{x_\alpha})_\alpha$ w^*-converges to some $\theta \in \mathcal{D}_\sigma \subset (J_\sigma^\perp \cap C_{\ell u}(G))^*$. Then

$$\langle R_x h_1, \theta \rangle = \lim_\alpha \langle R_x h_1, \bar{u}\,\delta_{x_\alpha} \rangle$$

$$= \lim_\alpha \bar{u}(x_\alpha) L_{x_\alpha^{-1}} h_1(x)$$

$$= \lim_\alpha \bar{u}(x_\alpha) L_{x_\alpha^{-1}} h_2(x)$$

$$= \langle R_x h_1, \theta \rangle$$

for every $x \in G$ which gives $\theta \circ h_1 = \theta \circ h_2$, that is,

$$\bar{s}_1(1)\theta \circ (s_1 \circ f) = \bar{s}_2(1)\theta \circ (s_2 \circ f).$$

Hence for any $\mu \in \mathcal{D}_\sigma$, we have

$$\bar{s}_1(1)\,\langle s_1 \circ f, \theta \circ \mu \rangle = \bar{s}_2(1)\,\langle s_2 \circ f, \theta \circ \mu \rangle.$$

Since u is a character, we have $\Pi_\sigma \circ (\bar{u}\,\delta_x) \subset \Pi_\sigma$ for $x \in G$ and hence Π_σ is a closed right ideal of $\mathcal{D}_\sigma$. So Π_σ contains an idempotent q such that $q \circ \mathcal{D}_\sigma$ is a minimal right ideal of $\mathcal{D}_\sigma$ (cf. [71]).

  In particular, we have $q \circ \mathcal{D}_\sigma = q \circ (\theta \circ \mathcal{D}_\sigma)$. Now, for any $t \in \Pi_\sigma$, we have

$$\bar{s}_1(1)\,\langle s_1 \circ f, t \rangle = \bar{s}_1(1)\,\langle f, s_1 \circ f \rangle$$

$$= \bar{s}_1(\mathbf{1}) \langle f, s_1 \circ q \circ t \rangle$$

$$= \bar{s}_1(\mathbf{1}) \langle s_1 \circ f, q \circ t \rangle$$

$$= \bar{s}_1(\mathbf{1}) \langle s_1 \circ f, q \circ \theta \circ t' \rangle \quad \text{for some} \quad t' \in \mathcal{D}_\sigma$$

$$= \bar{s}_1(\mathbf{1}) \langle \theta \circ q \circ (s_1 \circ f), t' \rangle$$

$$= \bar{s}_1(\mathbf{1}) \langle s_1 \circ f, \theta \circ (q \circ t') \rangle$$

$$= \bar{s}_2(\mathbf{1}) \langle s_2 \circ f, \theta \circ (q \circ t') \rangle$$

$$= \bar{s}_2(\mathbf{1}) \langle s_2 \circ f, t \rangle$$

which gives $h_1 = h_2$. $\qquad\qquad\qquad\qquad\qquad\qquad\qquad\qquad\qquad\Box$

As in the case of left σ-distal functions, we can define a *left distal function* $f \in J_\sigma^\perp \cap C_{\ell u}(G)$ by the left distal action of G on $\overline{L}_f$, given that $J_\sigma^\perp$ is left-translation invariant. If $\Pi_\sigma \cap \{\bar{u}\delta_x : x \in G\} \neq \emptyset$, then we have $\Pi_\sigma = \{\bar{u}\delta_x : x \in G\}^-$ by Proposition 2.2.21, and $\overline{L}_f = L_f^\sigma$ for if $g = \lim_\alpha L_{x_\alpha^{-1}} f$ pointwise on G, then a subnet of $(\bar{u}\delta_{x_\alpha})$ w^*-converges to some $\theta \in \Pi_\sigma$ and $g = \bar{\theta}(\mathbf{1})(\theta \circ f) \in L_f^\sigma$. Therefore, in this case, the left distal functions and the left σ-distal functions in $J_\sigma^\perp \cap C_{\ell u}(G)$ coincide.

Theorem 2.5.5. *Let $J_\sigma^\perp$ be translation invariant with identity u which is a continuous character. The following conditions are equivalent:*

(i) Π_σ *is isomorphic to a direct product of a right zero semigroup and a group;*

(ii) Π_σ *has a left identity and every $f \in J_\sigma^\perp \cap C_{\ell u}(G)$ is left σ-distal.*

Proof. (i) $\Rightarrow$ (ii). Let $\Pi_\sigma \simeq E \times H$ where E is a right zero semigroup and H is a group with identity θ. Then the proof of Theorem 2.5.4 implies that every $f \in J_\sigma^\perp \cap C_{\ell u}(G)$ is left σ-distal. Pick any $s \in E$. Then (s, θ) is a left identity as $(s, \theta)(t, w) = (st, \theta w) = (t, w)$ for $(t, w) \in E \times H$.

(ii) $\Rightarrow$ (i). Analogous to Theorem 2.5.2(i). $\qquad\qquad\qquad\qquad\qquad\Box$

2.6. Transitive group actions on Poisson spaces

Given a spread out probability measure σ on G, Azencott [5] has studied the Poisson space Π_σ in detail and in particular, he has shown that given two such non-degenerate measures σ_1 and σ_2, their Poisson spaces Π_{σ_1} and Π_{σ_2}

are equivariantly homeomorphic if G acts transitively on them. In this section, we supplement Azencott's results by applying our previous results on Π_σ to the case when σ is a probability measure. We show, for instance, that the above homeomorphism of Azencott also preserves the semigroup structure.

We assume, *throughout this section*, that $J_\sigma^\perp$ is left-translation invariant and that σ is an absolutely continuous probability measure with $u = 1$. Then every $f \in J_\sigma^\perp \cap C_{\ell u}(G) = H(\sigma)$ has the Poisson representation

$$f(x) = \int_{\Pi_\sigma} \widehat{f}(x^{-1} \cdot s) d\nu_\sigma(s) \quad (x \in G)$$

where $f \in H(\sigma) \mapsto \widehat{f} \in C(\Pi_\sigma)$ is the Gelfand map which is a surjective linear isometry and we denote its inverse map by $\widehat{f} \mapsto \widehat{f} \cdot \nu_\sigma$. The group action is given by $x \cdot t = R_{x^{-1}}^*(t)$ for $(x, t) \in G \times \Pi_\sigma$. We also define $x \cdot \widehat{f} \in C(\Pi_\sigma)$ by $(x \cdot \widehat{f})(t) = \widehat{f}(x \cdot t)$ for $t \in \Pi_\sigma$.

Theorem 2.6.1. *Let σ_1 and σ_2 be absolutely continuous probability measures on a second countable group G such that $\bigcup\limits_{n=1}^{\infty} \mathrm{supp}\, \sigma_1^n = \bigcup\limits_{n=1}^{\infty} \mathrm{supp}\, \sigma_2^n$. Then Π_{σ_1} and Π_{σ_2} are isomorphic as compact left topological semigroups if G acts transitively on them.*

Proof. By [5, Théorème II.4], there exists a homeomorphism $\rho : \Pi_{\sigma_1} \to \Pi_{\sigma_2}$ such that $\rho(x \cdot s) = x \cdot \rho(s)$ for $x \in G$ and $s \in \Pi_{\sigma_1}$. Then ρ induces an isometry, by composition, $\widehat{f} \in C(\Pi_{\sigma_2}) \mapsto \widehat{f} \circ \rho \in C(\Pi_{\sigma_1})$ which we will denote by ρ^*. Let $s, t \in \Pi_{\sigma_1}$. We need to show $\rho(s \circ t) = \rho(s) \circ \rho(t)$. Let $f \in H(\sigma_2)$. Then we have $\langle f, \rho(s) \circ \rho(t) \rangle = \langle \rho(s) \circ f, \rho(t) \rangle$ where, for $x \in G$, we have

$$(\rho(s) \circ f)(x) = \langle R_x f, \rho(s) \rangle$$
$$= \langle f, R_x^*(\rho(s)) \rangle$$
$$= \langle f, x^{-1} \cdot \rho(s) \rangle$$
$$= \langle \widehat{f}, \rho(x^{-1} \cdot s) \rangle = \rho^*(\widehat{f})(x^{-1} \cdot s).$$

Write $h = \rho^*(\widehat{f}) \cdot \nu_{\sigma_1} \in H(\sigma_1)$. Then $\rho^*(\widehat{f})(x^{-1} \cdot s) = \langle h, x^{-1} \cdot s \rangle = \langle h, R_x^*(s) \rangle = \langle R_x h, s \rangle = (s \circ h)(x)$ so $s \circ h = \rho(x) \circ f \in H(\sigma_2) \cap H(\sigma_1)$. The group action induces the following action on the Poisson measure

$$(x \cdot \nu_{\sigma_2})(\widehat{f}) = \int_{\Pi_{\sigma_2}} \widehat{f}(x^{-1} \cdot s) d\nu_{\sigma_2}(s).$$

By [5, Théorème II.4 and Lemme II.7], there is a sequence (x_n) in G such that $t = w^* - \lim_n x_n \cdot \nu_{\sigma_1}$ and $\rho(t) = w^* - \lim_n x_n \cdot \nu_{\sigma_2}$. It follows that

$$\langle f, \rho(s) \circ \rho(t) \rangle = \langle s_0 h, \rho(t) \rangle$$

$$= \lim_n \langle s \circ h, x_n \cdot \nu_{\sigma_2} \rangle$$

$$= \lim_n \langle s \circ h \cdot \nu_{\sigma_2}, x_n \rangle$$

$$= \lim_n \langle s \circ h, x_n \rangle$$

$$= \lim_n \langle s \circ h \cdot \nu_{\sigma_1}, x_n \rangle$$

$$= \langle s \circ h, t \rangle = \widehat{h}(s \circ t)$$

$$= \rho^*(\widehat{f})(s \circ t) = \langle f, \rho(s \circ t) \rangle.$$

Hence $\rho(s \circ t) = \rho(s) \circ \rho(t)$. $\qquad\qquad\qquad\qquad\qquad\qquad\square$

Proposition 2.6.2. *Let σ be a probability measure and let G act transitively on Π_σ^w. Then Π_σ^w is a compact topological group and $J_\sigma^\perp \cap \mathcal{WAP}(G) = J_\sigma^\perp \cap \mathcal{AP}(G)$.*

Proof. As in the proof of Lemma 2.5.3, one can show that Π_σ^w has no proper right ideal and therefore coincides with its minimal two-sided ideal K which is a compact topological group with identity θ say. Let $f \in J_\sigma^\perp \cap \mathcal{WAP}(G) \simeq C(\Pi_\sigma^w)$. Then for $s \in \Pi_\sigma^w$, we have

$$\langle R_x f, s \rangle = \langle f, R_x^* s \rangle$$

$$= \langle f, x^{-1} \cdot s \rangle$$

$$= \langle f, x^{-1} \cdot (s \circ \theta) \rangle$$

$$= \langle f, s \circ (x^{-1} \cdot \theta) \rangle = f_{x^{-1} \cdot \theta}(s)$$

where, as in the proof of Theorem 2.4.8, we define $f_t(s) = \langle f, s \circ t \rangle$ for $t \in \Pi_\sigma^w$. Therefore we have

$$\{R_x f : x \in G\} \subset \{f_t : t \in \Pi_\sigma^w\}$$

and the latter is norm compact in $C(\Pi_\sigma^w)$ by continuity of the map $t \in \Pi_\sigma^w \mapsto f_t \in C(\Pi_\sigma^w)$. So $f \in AP(G)$.   □

Corollary 2.6.3. *Let G be a semi-simple connected Lie group and let σ be an absolutely continuous probability measure on G such that $J_\sigma^\perp$ is left-translation invariant. Then every weakly periodic σ-harmonic function on G is constant.*

Proof. By [27], G acts transitively on Π_σ, and by [76; p. 184], we have $AP(G) = \mathbb{C}\mathbf{1}$.

Proposition 2.6.4. *Let σ_1 and σ_2 be absolutely continuous non-degenerate probability measures on a second countable group G such that $J_{\sigma_1}^\perp$ and $J_{\sigma_2}^\perp$ are left-translation invariant. Then the compact topological groups $\Pi_{\sigma_1}^a$ and $\Pi_{\sigma_2}^a$ are topologically isomorphic if G acts transitively on them.*

Proof. The arguments in the proof of [5; Théorème II.4] can be used to show that there exists an equivariant homeomorphism $\rho : \Pi_{\sigma_1}^a \to \Pi_{\sigma_2}^a$. The rest of the proof is similar to that of Theorem 2.6.1.   □

2.7. Examples

In contrast to the Choquet-Deny Theorem [11] asserting the absence of nonconstant bounded σ-harmonic functions on abelian groups for adapted probability measure σ, we have seen in Example 2.1.10 that nonconstant bounded σ-harmonic functions on $\mathbb{R}$ exist for complex adapted measures σ with $\|\sigma\| = 1$. Nevertheless, in this example, we still have $\dim J_\sigma^\perp = 1$, as shown below.

Lemma 2.7.1. *Let G be a locally compact abelian group and let σ be a complex adapted measure on G with $\|\sigma\| = 1 = \int_G \chi(-x) d\sigma(x)$ for some $\chi \in \widehat{G}$. Then $\dim J_\sigma^\perp = 1$.*

Proof. Let $d\mu(x) = \chi(-x) d\sigma(x)$. Then $\|\mu\| = 1 = \mu(G)$ and μ is an adapted probability measure on G. Therefore $\dim J_\sigma^\perp = 1$ since $f(x) \in J_\sigma^\perp$ if, and only if, $\chi(-x) f(x) \in J_\mu^\perp = \mathbb{C}\mathbf{1}$ which implies $J_\sigma^\perp = \mathbb{C}\chi$.   □

Let $C(\mathbb{R})$ denote the space of complex continuous functions on $\mathbb{R}$. One can describe $J_\sigma^\perp \cap C(\mathbb{R})$ completely, for $\sigma \in M(\mathbb{R})$ with compact support, using the results of Schwartz [73] for *mean periodic functions* on $\mathbb{R}$ which are the complex continuous functions f satisfying $\sigma * f = 0$ for some nonzero complex measure σ on $\mathbb{R}$ with compact support. We note that $J_\sigma^\perp = J_\sigma^\perp \cap C(\mathbb{R})$ if σ is absolutely continuous.

Lemma 2.7.2. *Let $\sigma \in M(\mathbb{R})$ have compact support. Then*

$$J_\sigma^\perp \cap C(\mathbb{R}) = \overline{\mathrm{lin}} \left(\left\{ e^{i\alpha x} : \alpha \in \mathbb{R} \text{ and } \int_{\mathbb{R}} e^{-i\alpha x} d\sigma(x) = 1 \right\} \cup \{0\} \right)$$

where the closure ' $-$ ' is taken in the uniform topology on compact sets in $\mathbb{R}$.

Proof. Let $\sigma_1 = \sigma - \delta_0$ where δ_0 is the point mass at 0. Let $f \in J_\sigma^\perp$. Then $\sigma_1 * f = 0$ and by Schwartz's result [73, Théorème 6], we have

$$f \in \overline{\mathrm{lin}} \{ p(x)e^{i\alpha x} : \alpha \in \mathbb{C} \text{ and } p(x)e^{i\alpha x} \in \tau(f) \}$$

where $\tau(f)$ denotes the closed linear span of $\{ R_x f : x \in \mathbb{R} \}$ in the uniform topology on compact sets in $\mathbb{R}$ and $p(x)$ is a polynomial. If $p(x)e^{i\alpha x} \in \tau(f)$, then $\sigma_1 * p(x)e^{i\alpha x} = 0$ which occurs if, and only if, either $p(x) = 0$ or $\widehat{\sigma}_1(\alpha) = \widehat{\sigma}_1'(\alpha) = \cdots = \widehat{\sigma}_1^{d(p)}(\alpha) = 0$ where $\widehat{\sigma}_1', \ldots, \widehat{\sigma}_1^{d(p)}$ are the derivatives of the Fourier transform $\widehat{\sigma}_1(\alpha) = \int_{\mathbb{R}} e^{-i\alpha x} d\sigma_1(x)$ $(\alpha \in \mathbb{C})$ and $d(p)$ is the degree of $p(x)$. But $p(x)e^{i\alpha x} \in L^\infty(\mathbb{R})$ implies that $d(p) = 0$ and $\alpha \in \mathbb{R}$. As $\widehat{\sigma} = \widehat{\sigma}_1 + \delta_0$, the result follows. $\qquad\square$

Let σ be a probability measure on $\mathbb{R}$. Then the subgroup S of $\mathbb{R}$ generated by supp σ is either dense in $\mathbb{R}$ or equal to $d\mathbb{Z}$ where d is the largest positive number such that $\frac{|x|}{d} \in \mathbb{Z}$ for all $x \in \operatorname{supp} \sigma$. By Choquet-Deny Theorem, every point in S is a period of any bounded continuous σ-harmonic function on $\mathbb{R}$. It follows that $\dim J_\sigma^\perp = 1$ or $\dim J_\sigma^\perp = \infty$. On the other hand, if σ is a probability measure on $\mathbb{Z}$ and $\sigma \neq \delta_0$, then the subgroup of $\mathbb{Z}$ generated by supp σ is equal to $d\mathbb{Z}$ where $d \in \mathbb{N}$ and it follows that $\dim J_\sigma^\perp = d$.

Example 2.7.3. Let $a, b \in \mathbb{R}$ with $a > b > 0$. Let $\sigma = \frac{1}{2}\delta_a + \frac{1}{2}\delta_b$. Then by the above remarks, we have $J_\sigma^\perp = \mathbb{C}\mathbf{1}$ if $\frac{a}{b}$ is irrational, and $\dim J_\sigma^\perp = \infty$ if $\frac{a}{b}$ is rational. In the latter case, we have

$$J_\sigma^\perp \cap C(\mathbb{R}) = \overline{\mathrm{lin}} \{ e^{i\alpha x} : \alpha \in \mathbb{R}, \ e^{-i\alpha a} + e^{-i\alpha b} = 2 \}$$

by Lemma 2.7.2. For instance, if $2a = 3b$ and $b = 1$, then $J_\sigma^\perp \cap C(\mathbb{R}) = \overline{\lin}\{e^{-4n\pi i x} : n \in \mathbb{Z}\}$.

Example 2.7.4. Let $\sigma \in M^1(\mathbb{R})$ be the measure

$$d\sigma(x) = \frac{1}{2\pi} e^{i\beta x} \chi_{[0,2\pi]} dx \qquad (\beta \in \mathbb{R}).$$

Then, for $\alpha \in \mathbb{R}$, we have

$$\widehat{\sigma}(\alpha) = \begin{cases} 1 & \text{if } \alpha = \beta \\ \frac{e^{2\pi i(\beta-\alpha)}-1}{2\pi i(\beta-\alpha)} \neq 1 & \text{if } \alpha \neq \beta. \end{cases}$$

Therefore $J_\sigma^\perp = \{ce^{i\beta x} : c \in \mathbb{C}\}$.

Remark 2.7.5. We note from Lemma 2.7.1 and Lemma 2.7.2 that if σ is a complex *adapted* measure on $\mathbb{R}$ with compact support and $\|\sigma\| = 1$, then there is at most one $\alpha \in \mathbb{R}$ such that $\widehat{\sigma}(\alpha) = 1$. This is not the case without adaptedness as shown in Example 2.7.3. For another example, the measure $\sigma = -i\delta_\pi$ has unit norm and $\widehat{\sigma}(n) = 1$ if n is an odd integer. The next example shows the existence of an adapted signed measure σ on $\mathbb{R}$ of unit norm such that $\widehat{\sigma}(\alpha) \neq 1$ for all $\alpha \in \mathbb{R}$ and hence $J_\sigma^\perp = \{0\}$.

Example 2.7.6. Let $d\sigma(x) = \varphi(x)dx$ be the measure on $\mathbb{R}$ given by

$$\varphi(x) = \begin{cases} -x & \text{if } -1 \leq x \leq 1 \\ 0 & \text{otherwise.} \end{cases}$$

Then $\|\sigma\| = 1$, $\sigma(\mathbb{R}) = \widehat{\sigma}(0) = 0$ and for $\alpha \in \mathbb{R}\backslash\{0\}$, $\widehat{\sigma}(\alpha) = \frac{2i}{\alpha}(\frac{\sin\alpha}{\alpha} - \cos\alpha) \neq 1$. By Lemma 2.7.2, we have $J_\sigma^\perp = \{0\}$ whereas $J_{|\sigma|}^\perp = \mathbb{C}1$ by Choquet-Deny Theorem.

Example 2.7.7. Let $\alpha_n = |\alpha_n|e^{i\theta_n}$ be such that $\sum_{n=1}^\infty |\alpha_n| = 1$ where the sequence (θ_n) is convergent and generates a dense subgroup of $\mathbb{R}$, for instance, $\alpha_n = \frac{6e^{\frac{1}{n}}}{n^2\pi^2}$. Then the bounded continuous solutions of the following functional equation

$$f(x) = \sum_{n=1}^\infty \alpha_n f(x - \theta_n)$$

are $\{ce^{ix} : c \in \mathbb{C}\}$. Indeed, let $\sigma = \sum \alpha_n \delta_{\theta_n}$. Then $\|\sigma\| = 1$ and σ is adapted with compact support. For $\alpha \in \mathbb{R}$, we have

$$\hat{\sigma}(\alpha) = \sum_{n=1}^{\infty} \alpha_n e^{-i\alpha\theta_n} = \sum_{n=1}^{\infty} |\alpha_n| e^{i(1-\alpha)\theta_n}.$$

Since $\hat{\sigma}(1) = 1$, we have $J_\sigma^\perp \cap C(\mathbb{R}) = \{ce^{ix} : c \in \mathbb{C}\}$.

We give an example of $\|\sigma\| > 1$.

Example 2.7.8. Let $\sigma \in M(\mathbb{R})$ be defined by $d\sigma(x) = \psi(x)dx + d\delta_0(x)$ where

$$\psi(x) = \begin{cases} \sin x & \text{for} \quad 0 \le x \le 2\pi \\ 0 & \text{otherwise.} \end{cases}$$

Then $\sigma(\mathbb{R}) = 1$, $\|\sigma\| = 5$ and $\hat{\sigma}(\alpha) = 1$ if, and only if, $\alpha \in \mathbb{Z}\backslash\{1\}$. By Lemma 2.7.2, we have $J_\sigma^\perp \cap C(\mathbb{R}) = \overline{\lim}\{e^{i\alpha x} : \alpha \in \mathbb{Z}\backslash\{1\}\}$.

Example 2.7.9. Given $\mu, \nu \in M_+(G)$, we denote their lattice infimum by $\mu \wedge \nu$. Let σ be a probability measure on G. A neighbourhood V of the identity is called σ -*admissible* if there exist $\ell \in \mathbb{N}$ and $\varepsilon > 0$ such that

$$\|(\sigma^\ell * \delta_a) \wedge (\sigma^\ell * \delta_b)\| > \varepsilon$$

for $a^{-1}b \in V$. Using martingale arguments, it has been shown in [17] recently that whenever $x^{-1}y$ belongs to such a neighbourhood V, then $f(x) = f(y)$ for every bounded σ-harmonic function f on G. If σ is nonsingular with its translates, but not necessarily absolutely continuous, then every compact invariant neighbourhood of the identity is σ -admissible [18]. It follows that, for such a measure σ on an [IN]-group G, every bounded σ-harmonic function is constant on each connected component of G, so $\dim J_\sigma^\perp \le \kappa$ where κ is the cardinality of the set of connected components of G. This result is also valid for non-degenerate absolutely continuous σ where we note that, if κ is finite or more generally, if G is almost connected which means that the quotient of G by the connected component G_e of the identity e is compact, then $J_\sigma^\perp = \mathbb{C}\mathbf{1}$ since each harmonic function on G induces naturally one on G/G_e and Lemma 2.1.6 applies.

Remark 2.7.10. Let G be a compact group and $\sigma \in M(G)$. As in the proof of Lemma 2.1.6, if $f \in J_\sigma^\perp$, then

$$f = \sum_{1 \in Sp(\hat{\sigma}(\pi))} \sum_{1 \le i,j \le \dim \pi} (\dim \pi) \widehat{f}(\pi)_{ji} \pi_{ij}$$

and it follows that $\dim J_\sigma^\perp \le \sum_{1 \in Sp(\hat{\sigma}(\pi))} (\dim \pi)^2$ where $\pi \in \widehat{G}$ and Sp denotes the spectrum.

3. Harmonic functionals on Fourier algebras

In this Chapter, we introduce and study a non-commutative analogue of harmonic functions, namely, the harmonic functionals on Fourier algebras. Let $B(G)$ be the Fourier-Stieltjes algebra of G and let $A(G)$ be the Fourier algebra which is a closed ideal of $B(G)$. The dual of $A(G)$ is the group von Neumann algebra $VN(G)$. Given $\sigma \in B(G)$ with $\|\sigma\| = 1$, the σ-*harmonic functionals* on $A(G)$ are defined to be the elements of the space $I_\sigma^\perp = \{T \in VN(G) : \sigma \cdot T = T\}$. We show that $I_\sigma^\perp$ is the range of a contractive projection P on $VN(G)$ and therefore it is a ternary Jordan algebra, that is, a Jordan triple system. We show further that it is a JW^*-algebra and that P is completely positive. The Jordan triple product for harmonic functionals with compact support is given by

$$2\{R, S, T\} = w^* - \lim_\alpha \mu_\alpha \cdot (RS^*T + TS^*R)$$

where μ_α belongs to the convex hull of $\{\sigma^n : n \geq 1\}$. We study the Jordan structure of $I_\sigma^\perp$ and the boundary components of its open unit ball which is a symmetric Banach manifold. We describe the Murray-von Neumann classification of $I_\sigma^\perp$ in terms of the convex geometry of its predual. We also study the annihilator I_σ of $I_\sigma^\perp$ in $A(G)$ and the Banach algebraic properties of the quotient $A(G)/I_\sigma$.

3.1. Fourier algebras

Our objective is to study the harmonic functionals on Fourier algebras. We first review some basic definitions and results concerning Fourier algebras in this section.

Let G be a locally compact group. We recall that the group C^*-algebra $C^*(G)$ of G is the completion of $L^1(G)$ with respect to the norm

$$\|f\|_c = \sup_\pi \{\|\pi(f)\|\}$$

where the supremum is taken over all ∗-representations $\pi : L^1(G) \to \mathcal{B}(H_\pi)$, the latter being the von Neumann algebra of bounded linear operators on the Hilbert space H_π. We will denote throughout by $\rho : G \to \mathcal{B}(L^2(G))$ the *left regular representation* of G :

$$\rho(x)h(y) = h(x^{-1}y) \quad (x, y \in G, h \in L^2(G)).$$

The unitary representation ρ can be extended to a ∗-representation of $L^1(G)$, also denoted by ρ :

$$\rho(f)h = f * h \quad (h \in L^2(G)).$$

The *reduced group C^*-algebra* $C_r^*(G)$ is the norm closure of $\rho(L^1(G))$ in $\mathcal{B}(L^2(G))$. We can further extend ρ to a representation $\tilde{\rho}$ of $C^*(G)$. Although ρ is injective on $L^1(G)$, $\tilde{\rho}$ need not be an isomorphism and in fact, $\tilde{\rho}$ is injective if, and only if, G is amenable. The group von Neumann algebra $VN(G)$ of G is the ultraweak closure of $\rho(L^1(G))$ in $\mathcal{B}(L^2(G))$ and is also the ultraweak closure of the linear span of $\rho(G)$ in $\mathcal{B}(L^2(G))$.

A function $\varphi : G \longrightarrow \mathbb{C}$ is called *positive definite* if

$$\sum_{i,j=1}^n \lambda_i \overline{\lambda}_j \varphi(x_i x_j^{-1}) \geq 0$$

for any $\lambda_1, \ldots, \lambda_n \in \mathbb{C}$ and $x_1, \ldots, x_n \in G$. A continuous positive definite function on G is of the form $\varphi(\cdot) = (\pi(\cdot)\eta, \eta)$, and vice versa, where $\{\pi, H\}$ is a continuous unitary representation of G and $\eta \in H$. Also, if $\varphi \in L^\infty(G)$, then φ is continuous positive definite if, and only if, it is a positive linear functional of $L^1(G)$, that is, $\langle \varphi, f^* * f \rangle \geq 0$ for all $f \in L^1(G)$. Let $P(G)$ be the subset of $C_b(G)$ consisting of all continuous positive definite functions on G and let $P^1(G) = \{\varphi \in P(G) : \varphi(e) = 1\}$. The linear span $B(G)$ of

$P^1(G)$ in $C_b(G)$ can be identified with the dual of $C^*(G)$ via the duality $\langle \cdot, \cdot \rangle : C^*(G) \times B(G) \to \mathbb{C}$ where

$$\langle f, \varphi \rangle = \int_G f(t)\varphi(t)d\lambda(t)$$

for $f \in L^1(G)$ and $\varphi \in B(G)$. In this duality, $P(G)$ is precisely the set of positive linear functionals on $C^*(G)$ [24; p. 192]. Moreover, $B(G)$ is a commutative semi-simple Banach algebra, called the *Fourier-Stieltjes algebra* of G, with pointwise multiplication and the dual norm

$$\|\varphi\| = \sup\left\{ \left| \int_G f(t)\varphi(t)d\lambda(t) \right| : f \in L^1(G), \|f\|_c \leq 1 \right\}$$

(see [24; Proposition 2.16]). Let $A(G)$ be the subspace of $B(G)$ consisting of all functions φ of the form

$$\varphi(x) = \big(\rho(x)h|k\big) \quad (x \in G, h, k \in L^2(G))$$

where $(\cdot|\cdot)$ denotes the inner product in $L^2(G)$. Then $A(G)$ is a closed ideal of $B(G)$ and is called the *Fourier algebra* of G. We note that $A(G) \subset C_0(G)$, $A(G) \cap C_c(G)$ is dense in $A(G)$, where $C_c(G)$ denotes the space of continuous functions on G with compact support, and the dual $A(G)^*$ is isometrically isomorphic to $VN(G)$ on which the ultraweak topology coincides with the w^*-topology $\sigma\big(VN(G), A(G)\big)$ [24; p. 210].

If G is abelian with dual group $\widehat{G}$, then $C^*(G) \cong C_0(\widehat{G}) \cong C_r^*(G)$ and the positive definite functions on G are precisely the Fourier transforms of positive bounded measures on $\widehat{G}$, so $B(G) = \big(M(\widehat{G})\big)^\wedge$. Also $A(G) = \big(L^1(\widehat{G})\big)^\wedge$ and $VN(G) \cong L^\infty(\widehat{G})$.

There is a natural action of $A(G)$ on $VN(G)$ given by

$$\langle \psi, \varphi \cdot T \rangle = \langle \varphi\psi, T \rangle \quad (\varphi, \psi \in A(G), T \in VN(G))$$

where $\langle \cdot, \cdot \rangle$ is the duality between $A(G)$ and $VN(G)$. Given $\sigma \in B(G)$ and $T \in VN(G)$, we define $\sigma \cdot T \in VN(G)$ by

$$\langle \varphi, \sigma \cdot T \rangle = \langle \sigma\varphi, T \rangle$$

for $\varphi \in A(G)$. We note that

$$\langle \varphi, \rho(x) \rangle = \varphi(x)$$

for $\varphi \in A(G)$ and $x \in G$. Thus $\rho(x)$ is the '*evaluation*' functional of $A(G)$ at x and is multiplicative.

As in [24; p. 225], we define the *support of* $T \in VN(G)$, supp T, to be the set of $x \in G$ such that $\rho(x)$ is the w^*-limit of a net $(\varphi_\alpha \cdot T)$ with $\varphi_\alpha \in A(G)$. We have, by [24; 4.4],

$$\operatorname{supp} T = \{x \in G : \varphi(x) = 0 \text{ for } \varphi \in A(G) \text{ with } \varphi \cdot T = 0\}.$$

By [24; 4.9], if supp $T = \{x_0\}$, then T is a complex multiple of $\rho(x_0)$.

We write $A(G) \cdot VN(G) = \{\varphi \cdot T : \varphi \in A(G), T \in VN(G)\}$ and $VN_c(G) = \{T \in VN(G) : \operatorname{supp} T \text{ is compact}\}$. Then the norm-closure $\overline{VN_c(G)}$ coincides with the closed linear span

$$\overline{\operatorname{span}} \left(A(G) \cdot VN(G) \right)$$

which is a C^*-algebra since

$$\operatorname{supp} (T_1 T_2) \subset (\operatorname{supp} T_1)(\operatorname{supp} T_2) \quad \text{and} \quad \operatorname{supp} T^* = (\operatorname{supp} T)^{-1}.$$

If G is amenable, then $\overline{\operatorname{span}} \left(A(G) \cdot VN(G) \right) = A(G) \cdot VN(G)$ by Cohen's factorization theorem [38; 32.22]. If G is abelian, then $\overline{\operatorname{span}} A(G) \cdot VN(G)$ identifies with the C^*-algebra of bounded uniformly continuous functions on $\widehat{G}$ and is denoted by $UCB(\widehat{G})$ in [33].

We denote by $P_\rho(G)$ the closure of $P(G) \cap C_c(G)$ in $B(G)$ with respect to the compact open topology. Let $B_\rho(G)$ be the linear span of $P_\rho(G)$ in $B(G)$. Then $B_\rho(G)$ is a closed ideal in $B(G)$ containing $A(G)$ and $\left(C_r^*(G) \right)^* = B_\rho(G)$ [24; 2.1 and 2.6].

3.2. Harmonic functionals and associated ideals

In this section, we introduce the concept of *harmonic functionals* as a noncommutative analogue of harmonic functions and we study the ideals associated with these functionals.

To motivate our definition, consider an abelian group G and a measure $\sigma \in M(\widehat{G})$ with J_σ equal to the norm closure of $\{\breve{\sigma} * f - f : f \in L^1(\widehat{G})\}$ as defined in Section 2.1. Let ν be the Fourier transform of $\breve{\sigma}$. Then J_σ has Fourier transform $\widehat{J}_\sigma$ equal to the norm closure of $\{\nu\varphi - \varphi : \varphi \in A(G)\}$. If we write $I_\nu = \widehat{J}_\sigma$ where $\nu \in B(G)$, then $I_\nu^\perp = \{T \in VN(G) : \nu \cdot T = T\}$ and identifies with $J_\sigma^\perp = \{h \in L^\infty(\widehat{G}) : \sigma * h = h\}$. This leads to the following natural definition.

Definition 3.2.1. Let G be a locally compact group and let $\sigma \in B(G)$. We define I_σ to be the norm-closure of $\{\sigma\varphi - \varphi : \varphi \in A(G)\}$ and we call the elements in

$$I_\sigma^\perp = \{T \in VN(G) : \sigma \cdot T = T\}$$

the *σ-harmonic functionals* on $A(G)$ or simply *harmonic functionals* if σ is understood.

By the above remarks, we can view $I_\sigma^\perp \subset VN(G)$ as a non-commutative analogue of $J_\sigma^\perp \subset L^\infty(G)$. We note that I_σ is a closed ideal in $A(G)$ and $(A(G)/I_\sigma)^* = I_\sigma^\perp$. Our main objective in the sequel is to study the structures of $A(G)/I_\sigma$ and $I_\sigma^\perp$. In this section, we study the ideal I_σ and the quotient algebra $A(G)/I_\sigma$.

We will always assume $\|\sigma\| \geq 1$ for if $\|\sigma\| < 1$, we have $I_\sigma^\perp = \{0\}$ as $T \in I_\sigma^\perp$ implies $\|T\| = \|\sigma \cdot T\| \leq \|\sigma\| \|T\|$. Let

$$B^1(G) = \{\sigma \in B(G) : \|\sigma\| = 1\}.$$

Lemma 3.2.2. *Let G be amenable and $\sigma \in B^1(G)$. Then I_σ has a bounded approximate identity.*

Proof. By amenability, $A(G)$ has a bounded approximate identity [58] $\{\varphi_\alpha\}_{\alpha \in \Lambda}$ which can be chosen such that $\varphi_\alpha \geq 0$ and $\varphi_\alpha(e) = 1$ for all $\alpha \in \Lambda$ [53]. Let $\sigma_n = \frac{1}{n}\sum_{k=1}^{n}\sigma^k$ for $n = 1, 2, \ldots$. Then $\|\sigma_n\| \leq 1$. We show that $\{\varphi_\alpha - \sigma_n\varphi_\alpha\}_{\alpha,n}$ is a bounded approximate identity in I_σ. First $\sigma_n\varphi_\alpha - \varphi_\alpha \in I_\sigma$ as $A(G)$ is an ideal in $B(G)$.

Given $\psi \in A(G)$, we have

$$\|(\sigma\psi - \psi) - (\sigma\psi - \psi)(\varphi_\alpha - \sigma_n\varphi_\alpha)\|$$
$$= \|(\psi\varphi_\alpha - \psi) + \sigma(\psi - \psi\varphi_\alpha) + (\sigma\sigma_n - \sigma_n)\psi\varphi_\alpha\|$$
$$\leq \|\psi - \psi\varphi_\alpha\| + \|\sigma(\psi - \psi\varphi_\alpha)\| + \|\sigma\sigma_n - \sigma_n\| \|\psi\varphi_\alpha\|$$
$$\leq 2\|\psi - \psi\varphi_\alpha\| + \frac{1}{n}\|\psi\varphi_\alpha\|$$

which tends to 0 as $n, \alpha \to \infty$ by boundedness of $\{\varphi_\alpha\}$. □

Let $A(G)_0 = \{\varphi \in A(G) : \varphi(e) = 0\}$. Then $A(G)_0$ is a closed ideal in $A(G)$ with co-dimension one. Further, $A(G)_0$ has a bounded approximate identity if, and only if, G is amenable [52; Corollary 4.11].

Lemma 3.2.3. *Let* G *be a locally compact group. Then* $A(G)_0$ *is the closed linear span of* $\{I_\sigma : \sigma \in A(G) \cap P^1(G)\}$.

Proof. Let $\mathcal{I}$ be the closed linear span of $\{I_\sigma : \sigma \in A(G) \cap P^1(G)\}$. Then $\mathcal{I} \subset A(G)_0$. It suffices to show that $\mathcal{I}$ has co-dimension one, equivalently, that $\mathcal{I}^\perp = \mathbb{C}1$ where 1 is the identity of $VN(G)$. Let $T \in \mathcal{I}^\perp$. Then $\sigma \cdot T = T$ for all $\sigma \in A(G) \cap P^1(G)$. Let (σ_α) be a net in $A(G) \cap P^1(G)$ such that $\operatorname{supp} \sigma_\alpha \downarrow \{e\}$. Then $\operatorname{supp}(\sigma_\alpha \cdot T) \subset \operatorname{supp} \sigma_\alpha$ gives $\operatorname{supp} T = \{e\}$. Therefore T is a complex multiple of $\rho(e) = 1$. $\qquad\square$

Lemma 3.2.4. *Let* G *be first countable. Then* $A(G)_0 = I_\sigma$ *for some* $\sigma \in A(G) \cap P^1(G)$.

Proof. Let (φ_α) be a net in $A(G) \cap P^1(G)$ such that $\|\sigma\varphi_\alpha - \varphi_\alpha\| \to 0$ for each $\sigma \in A(G) \cap P^1(G)$ [67]. Then $\|(\sigma\varphi - \varphi) \cdot \varphi_\alpha\| = \|\varphi(\sigma\varphi_\alpha - \varphi_\alpha)\| \to 0$ for $\sigma \in A(G) \cap P^1(G)$ and $\varphi \in A(G)$. Let $\varepsilon > 0$ and $\varphi_1, \ldots, \varphi_n \in A(G)_0$. By the above and Lemma 3.2.3, there exists $\varphi_\beta \in A(G) \cap P^1(G)$ such that $\|\varphi_i\varphi_\beta\| < \varepsilon$ for $i = 1, \ldots, n$. It follows from $\varphi_\beta\varphi_i - \varphi_i \in I_{\varphi_\beta}$ that

$$d(\varphi_i, I_{\varphi_\beta}) = \inf\{\|\varphi_i - \psi\| : \psi \in I_{\varphi_\beta}\} < \varepsilon$$

for $i = 1, \ldots, n$. Since G is first countable, $A(G)$ is norm separable [29; Corollary 6.9]. So the conditions of Lemma 1.1 in [83] are satisfied and by Remark 3 in [83, p. 210], there exists $\sigma \in A(G) \cap P^1(G)$ such that $I_\sigma = A(G)_0$. $\qquad\square$

Theorem 3.2.5. *A first countable group* G *is amenable if, and only if,* I_σ *has a bounded approximate identity for every* $\sigma \in B^1(G)$.

Proof. The necessity has been proved in Lemma 3.2.2. The converse follows from Lemma 3.2.4 and the previous remark that amenability of G is equivalent to $A(G)_0$ having a bounded approximate identity. $\qquad\square$

We are going to prove a non-commutative analogue of Proposition 2.1.3. We prove a simple lemma first.

Lemma 3.2.6. *Let* $\sigma \in P^1(G)$. *The following conditions are equivalent:*

(i) $I_\sigma = A(G)_0$;
(ii) $I_\sigma^\perp = \mathbb{C}1$;
(iii) *For* $x \in G$, $\sigma(x) = 1$ *implies* $x = e$.

Proof. (i) $\Longleftrightarrow$ (ii). This is clear from $I_\sigma \subset A(G)_0$ and $A(G)_0^\perp = \mathbb{C}1$.

(ii) $\Longrightarrow$ (iii). Let $\sigma(x) = 1$. Then $\sigma \cdot \rho(x) = \rho(x)$ since $\langle \varphi, \sigma \cdot \rho(x) \rangle = \langle \sigma\varphi, \rho(x) \rangle = \sigma(x)\varphi(x) = \varphi(x) = \langle \varphi, \rho(x) \rangle$ for all $\varphi \in A(G)$. Hence $\rho(x) \in I_\sigma^\perp = \mathbb{C}1$ gives $x = e$.

(iii) $\Longrightarrow$ (ii). Let $T \in I_\sigma^\perp$. We show that $\operatorname{supp} T = \{e\}$. Otherwise, there exists $x \in \operatorname{supp} T\backslash\{e\}$ and hence a net (φ_α) in $A(G)$ such that $(\varphi_\alpha \cdot T)$ w^*-converges to $\rho(x)$.

Choose $\psi \in A(G)$ with $\psi(x) = 1$. Then $\sigma \cdot (\varphi_\alpha \cdot T) = \varphi_\alpha \cdot T$ for all α and

$$1 = \psi(x) = \langle \psi, \rho(x) \rangle$$
$$= \lim_\alpha \langle \psi, \varphi_\alpha \cdot T \rangle$$
$$= \lim_\alpha \langle \psi, \sigma \cdot (\varphi_\alpha \cdot T) \rangle$$
$$= \langle \psi, \sigma \cdot \rho(x) \rangle = \sigma(x)\psi(x) = \sigma(x) \neq 1$$

which is a contradiction. So $\operatorname{supp} T = \{e\}$ and $T \in \mathbb{C}1$. $\qquad\square$

The following result should be compared with Proposition 2.1.3.

Proposition 3.2.7. *Let G be a locally compact group. The following conditions are equivalent:*

(i) *There exists $\sigma \in P^1(G)$ such that $I_\sigma^\perp = \mathbb{C}1$;*

(ii) *G is first countable.*

In the above case, σ can even be chosen from $A(G)$.

Proof. By Lemma 3.2.4, we have (ii) $\Longrightarrow$ (i).

(i) $\Longrightarrow$ (ii). By Lemma 3.2.6, there exists $\sigma \in P^1(G) \subset C_{ru}(G)$ such that $\sigma(x) \neq 1$ for $x \neq e$. We first note that a net (x_α) in G converges to e if, and only if, $\sigma(x_\alpha) \to \sigma(e)$. Indeed, the latter implies that (x_α) is eventually in a compact neighbourhood of e, and if $x_\alpha \nrightarrow e$, there is a subnet (x_β) converging to some $x \neq e$. But $\sigma(x) = \lim_\beta \sigma(x_\beta) = \sigma(e) = 1$ which is impossible.

Let C be a compact neighbourhood of e and let $K = \{L_a\sigma : a \in C\}$ be the left translations of σ by C. Then K is compact in the sup-norm topology of $C_{ru}(G)$ since the map $a \in C \mapsto L_a\sigma \in C_{ru}(G)$ is continuous. It follows that K has a norm-dense sequence (ψ_n) and that a net $x_\alpha \to x$ in C if, and only if, $\psi_n(x_\alpha) \to \psi_n(x)$ for all n. So C is metrizable and G is first countable. $\qquad\square$

Now we are going to study the quotient algebra $A(G)/I_\sigma$. We first find sufficient conditions for $A(G)/I_\sigma$ to be a *Fourier algebra*, more precisely, to be isomorphic to the Fourier algebra of a locally compact group. Given an ideal I in $A(G)$, we define its *zero set* to be the closed set

$$Z(I) = \{x \in G : \varphi(x) = 0 \text{ for all } \varphi \in I\}.$$

Lemma 3.2.8. *For every $\sigma \in B(G)$, we have $Z(I_\sigma) = \{x \in G : \sigma(x) = 1\}$. In particular, if $\sigma \in P^1(G)$, then $Z(I_\sigma)$ is a closed subgroup of G.*

Proof. If $\sigma(x) = 1$, then $(\sigma\varphi - \varphi)(x) = 0$ for all $\varphi \in A(G)$ and so $x \in Z(I_\sigma)$. If $\sigma(x) \neq 1$, pick $\psi \in A(G)$ such that $\psi(x) = 1$. Then $(\sigma\psi - \psi)(x) \neq 0$ and $x \notin Z(I_\sigma)$. If $\sigma \in P^1(G)$, then $\{x \in G : \sigma(x) = 1\}$ is a subgroup of G by [38; 32.7]. $\qquad\square$

Remark 3.2.9. We note that $x \in Z(I_\sigma)$ if, and only if, $\rho(x) \in I_\sigma^\perp$. Indeed, given $\rho(x) \in I_\sigma^\perp$, pick $\psi \in A(G)$ with $\psi(x) = 1$, then $\langle \sigma\psi - \psi, \rho(x) \rangle = 0$ implies $\sigma(x) = 1$. The converse has been shown in the proof of Lemma 3.2.6 (ii) $\Longrightarrow$ (iii).

Notation. For $\sigma \in B(G)$, we write $Z_\sigma = Z(I_\sigma) = \sigma^{-1}\{1\}$.

Given a subset S of $VN(G)$, we denote by S' its commutant in $VN(G)$.

Proposition 3.2.10. *Let $\sigma \in B(G)$. The following conditions are equivalent:*

(i) $Z(I_\sigma)$ *is a subgroup of G;*
(ii) $I_\sigma^\perp$ *is a unital $*$-subalgebra of $VN(G)$.*

In the above case, $I_\sigma^\perp = \{\rho(x) : x \in Z(I_\sigma)\}''$ is the von Neumann algebra generated by $\rho(Z(I_\sigma))$ and further, there is an isometric algebra isomorphism Ψ from $A(G)/I_\sigma$ onto the Fourier algebra $A(Z(I_\sigma))$ such that $\Psi\{\varphi + I_\sigma : \sigma \in A(G) \cap P(G)\} = A(Z(I_\sigma)) \cap P(Z(I_\sigma))$.

Proof. (i) $\Longrightarrow$ (ii). Let $T \in VN(G)$. By [75], we have $T \in \{\rho(x) : x \in Z(I_\sigma)\}''$ if, and only if, $\operatorname{supp} T \subset Z(I_\sigma)$. By the double commutant theorem, $\{\rho(x) : x \in Z(I_\sigma)\}''$ is the w^*-closed linear span of $\rho(Z(I_\sigma))$ which is contained in $I_\sigma^\perp$ by Remark 3.2.9. We show that they are equal. Let $T \in I_\sigma^\perp$. We show $\operatorname{supp} T \subset Z(I_\sigma)$. Let $x \in \operatorname{supp} T$. Then there is a net (φ_α) in $A(G)$ such that $(\varphi_\alpha \cdot T)$ w^*-converges to $\rho(x)$. As $\varphi_\alpha \cdot T \in I_\sigma^\perp$, we have $\rho(x) \in I_\sigma^\perp$ and $x \in Z(I_\sigma)$.

(ii) $\Longrightarrow$ (i). Since $I_\sigma^\perp \neq \{0\}$, we have $I_\sigma \neq A(G)$ and so $Z(I_\sigma) \neq \emptyset$ [24;

3.38]. Given $x, y \in Z(I_\sigma)$, we have $\rho(xy^{-1}) = \rho(x)\rho(y)^* \in I_\sigma^\perp$ and hence $xy^{-1} \in Z(I_\sigma)$.

Now let $H = Z(I_\sigma)$ be a subgroup of G and define $\Psi : A(G)/I_\sigma \to A(H)$ by restriction:

$$\Psi(\varphi + I_\sigma) = \varphi\big|_H.$$

Since $I_\sigma^\perp = \{\rho(x) : x \in H\}''$, given $\varphi \in A(G)$, we have $\varphi \in I_\sigma$ if, and only if, $\varphi\big|_H = 0$ by the bi-polar rule. By [37, Theorem 1b], every $\psi \in A(H)$ has a norm-preserving extension to $\varphi \in A(G)$. The above two remarks imply that Ψ is an isometric isomorphism. Finally, given $\varphi \in A(G) \cap P(G)$, clearly $\varphi\big|_H \in A(H) \cap P(H)$. Conversely, for $\psi \in A(H) \cap P(H)$, it has a norm-preserving extension to $\varphi \in A(G)$. Then $\Psi(\varphi + I_\sigma) = \psi$ and $\|\varphi\| = \|\psi\| = \psi(e) = \varphi(e)$ implies $\varphi \in P(G)$. $\qquad\qquad\square$

Remark 3.2.11. We note that in the above theorem, $\{\varphi + I_\sigma : \varphi \in P(G) \cap A(G)\}$ is the set of normal positive functionals of the von Neumann algebra $I_\sigma^\perp$. In contrast to Corollary 2.2.4, if $\sigma \in P^1(G)$, then $I_\sigma^\perp$ is always a subalgebra of $VN(G)$.

We next consider the case in which the zero set Z_σ has a group structure but need not be a subgroup of G. We make use of [81] to determine when $A(G)/I_\sigma$ is isometrically isomorphic to the Fourier algebra $A(Z_\sigma)$.

We recall that the *spectrum* of $A(G)$, the set of nonzero multiplicative linear functionals on $A(G)$ equipped with the w^*-topology, is homeomorphic to G via the homeomorphism $x \in G \mapsto \rho(x)$ (cf. [24; 3.34]). It follows from this and $(A(G)/I_\sigma)^* = I_\sigma^\perp$ that the spectrum of $A(G)/I_\sigma$ is homeomorphic to the zero set Z_σ and the Gelfand map on $A(G)/I_\sigma$ is a norm-decreasing algebra homomorphism from $A(G)/I_\sigma$ onto a subalgebra of $C_0(Z_\sigma)$. The composite of the Gelfand map with the quotient map

$$A(G) \to A(G)/I_\sigma \to C_0(Z_\sigma)$$

is simply the restriction map $\varphi \in A(G) \mapsto \varphi\big|_{Z_\sigma} \in C_0(Z_\sigma)$.

Now we are ready to give necessary and sufficient conditions for $A(G)/I_\sigma$ to be a Fourier algebra of a group. Clearly a necessary condition is that the zero set $Z_\sigma = Z(I_\sigma)$ should have a group structure since it is the spectrum of $A(G)/I_\sigma$.

Theorem 3.2.12. *Let $\sigma \in B(G)$ and $\sigma(e) = 1$. The following conditions are equivalent:*

(i) $A(G)/I_\sigma$ *is isometrically algebraically isomorphic to a Fourier algebra of a locally compact group;*

(ii) *the following two conditions hold:*

(α) $I_\sigma^\perp$ *is a von Neumann algebra in some product* $\times$ *and involution* #;

(β) $\rho(Z_\sigma)$ *is a group of unitaries in* $(I_\sigma^\perp, \times, \#)$ *with identity* $\rho(e)$ *and* $I_\sigma^\perp$ *is the* w^*-*closed linear span of* $\rho(Z_\sigma)$.

Proof. We note that $\sigma(e) = 1$ implies $e \in Z_\sigma$ and $\rho(e) \in I_\sigma^\perp$.

(i) $\Longrightarrow$ (ii). This is clear for if $\Psi : A(G)/I_\sigma \to A(H)$ is an isometric isomorphism onto the Fourier algebra $A(H)$ of a group H, then the dual map $\Psi^* : VN(H) \to I_\sigma^\perp$ induces a von Neumann algebraic structure on $I_\sigma^\perp$ and maps the spectrum H of $A(H)$ onto the spectrum $\rho(Z_\sigma)$ of $A(G)/I_\sigma$.

(ii) $\Longrightarrow$ (i). Let $(I_\sigma^\perp, \times, \#)$ be a von Neumann algebra with identity $\mathbf{1} = \rho(e)$ and predual $A(G)/I_\sigma$. Then $\rho(Z_\sigma)$ is a locally compact group in the w^*-topology. We show that the algebra $A(G)/I_\sigma$ satisfies the conditions (i)-(vi) in [81; Theorem 6] and hence $A(G)/I_\sigma$ is isometrically isomorphic to the Fourier algebra $A(\rho(Z_\sigma))$. These conditions are shown below as (i')-(vi').

(i') Let $P = \{\varphi + I_\sigma \in A(G)/I_\sigma : \langle \varphi + I_\sigma, \rho(e) \rangle = \|\varphi + I_\sigma\|\}$. Then $A(G)/I_\sigma$ is the linear span of P. Indeed, $A(G)/I_\sigma$ is the linear span of the set of positive normal functionals of $(I_\sigma^\perp, \times, \#)$ which is just P.

(ii') Write $[\varphi] = \varphi + I_\sigma$ for $\varphi \in A(G)$. Let Aut denote the set of isometric algebra automorphisms of $A(G)/I_\sigma$ and let

$$T = \{\Lambda \in \text{Aut} : \| [\varphi] - e^{i\theta} \Lambda [\varphi] \|^2 + \| [\varphi] + e^{i\theta} \Lambda [\varphi] \|^2 \le 4\| [\varphi] \|^2$$

$$\forall [\varphi] \in A(G)/I_\sigma \quad \text{and} \quad \theta \in \mathbb{R}\}.$$

For $z \in Z_\sigma$, define $L_z : A(G)/I_\sigma \to A(G)/I_\sigma$ by

$$\langle L_z[\varphi], T \rangle = \langle \varphi, \rho(z) \times T \rangle \qquad ([\varphi] \in A(G)/I_\sigma, T \in I_\sigma^\perp).$$

Then the dual map $L_z^* : I_\sigma^\perp \to I_\sigma^\perp$ is just the left translation (w.r.t. $\times$) by $\rho(z)$ and is therefore an isometry. So L_z is an isometry. For $[\varphi_1], [\varphi_2] \in A(G)/I_\sigma$ and $x \in Z_\sigma$, we have

$$\langle L_z([\varphi_1][\varphi_2]), \rho(x) \rangle = \langle L_z[\varphi_1\varphi_2], \rho(x) \rangle$$

$$= \langle \varphi_1\varphi_2, \rho(z) \times \rho(x) \rangle$$

$$= \langle \varphi_1\varphi_2, \rho(y) \rangle \quad ((\rho(Z_\sigma), \times) \text{ is a group})$$

$$= \varphi_1(y)\varphi_2(y)$$

$$= \langle \varphi_1, \rho(z) \times \rho(x) \rangle \langle \varphi_2, \rho(z) \times \rho(x) \rangle$$

$$= \langle L_z[\varphi_1], \rho(x) \rangle \langle L_z[\varphi_2], \rho(x) \rangle.$$

Since $I_\sigma^\perp$ is the w*-closed linear span of $\rho(Z_\sigma)$, we have shown $L_z \in \text{Aut}$. We next show that $L_z \in T$. First, let $[\varphi] \in A(G)/I_\sigma$ be a positive (normal) state of the von Neumann algebra $(I_\sigma^\perp, \times, \#)$ and let

$$\langle [\varphi], T \rangle = (\pi_\varphi(T)\xi_\varphi \,|\, \xi_\varphi) \quad (T \in I_\sigma^\perp)$$

where $\pi_\varphi : I_\sigma^\perp \to B(H_\varphi)$ is the GNS-representation induced by $[\varphi]$ and $\xi_\varphi \in H_\varphi$ is a cyclic vector with $\|\xi_\varphi\| = \langle [\varphi], 1 \rangle = 1$.

Let $S, T \in I_\sigma^\perp$ with $\|S\|, \|T\| \leq 1$. Then

$$|\langle [\varphi] - e^{i\theta} L_z[\varphi], S \rangle|^2 + |\langle [\varphi] + e^{i\theta} L_z[\varphi], T \rangle|^2$$

$$= \left| (\pi_\varphi(S)\xi_\varphi | \xi_\varphi) - e^{i\theta}(\pi_\varphi(\rho(z) \times S)\xi_\varphi | \xi_\varphi) \right|^2$$

$$+ \left| (\pi_\varphi(T)\xi_\varphi | \xi_\varphi) + e^{i\theta}(\pi_\varphi(\rho(z) \times T)\xi_\varphi | \xi_\varphi) \right|^2$$

$$= \left| (\pi_\varphi(S)\xi_\varphi | \xi_\varphi - e^{i\theta}\pi_\varphi(\rho(z)\#)\xi_\varphi) \right|^2$$

$$+ \left| (\pi_\varphi(T)\xi_\varphi | \xi_\varphi + e^{i\theta}\pi_\varphi(\rho(z)\#)\xi_\varphi) \right|^2$$

$$\leq \|\pi_\varphi(S)\xi_\varphi\|^2 \|\xi_\varphi - e^{i\theta}\pi_\varphi(\rho(z)\#)\xi_\varphi\|^2$$

$$+ \|\pi_\varphi(T)\xi_\varphi\|^2 \|\xi_\varphi + e^{i\theta}\pi_\varphi(\rho(z)\#)\xi_\varphi\|^2$$

$$\leq \|\xi_\varphi - e^{i\theta}\pi_\varphi(\rho(z)\#)\xi_\varphi\|^2 + \|\xi_\varphi + e^{i\theta}\pi_\varphi(\rho(z)\#)\xi_\varphi\|^2$$

$$= 2\|\xi_\varphi\|^2 + 2\|e^{i\theta}\pi_\varphi(\rho(z)\#)\xi_\varphi\|^2 \leq 4$$

by the parallelogram law in H_φ.

Next, for any $[\psi] \in A(G)/I_\sigma$, let $[\psi] = [\varphi]V$ be the polar decomposition where V is a partial isometry in the von Neumann algebra $(I_\sigma^\perp, \times, \#)$ and $[\varphi]$ is a positive functional of $I_\sigma^\perp$ with

$$\langle [\psi], T \rangle = \langle [\varphi], T \times V \rangle$$

for $T \in I_\sigma^\perp$. Then we have

$$\| [\psi] - e^{i\theta} L_z [\psi] \|^2 + \| [\psi] + e^{i\theta} L_z [\psi] \|^2$$

$$= \| [\varphi]V - e^{i\theta} L_z [\varphi]V \|^2 + \| [\varphi]V + e^{i\theta} L_z [\varphi]V \|^2$$

$$= \| ([\varphi] - e^{i\theta} L_z [\varphi])V \|^2 + \| ([\varphi] + e^{i\theta} L_z [\varphi])V \|^2$$

$$\leq \| [\varphi] - e^{i\theta} L_z [\varphi] \|^2 + \| [\varphi] + e^{i\theta} L_z [\varphi] \|^2 \leq 4\| [\varphi] \|^2.$$

Hence we have shown that $L_z \in T$. Let G_σ be a maximal group in T containing $\{L_z : z \in Z_\sigma\}$. Then we can define a map $\natural : G_\sigma \to \rho(Z_\sigma)$ by

$$\natural(\Lambda) = \Lambda^*(1)$$

where $\Lambda^*(1)$ is multiplicative on $A(G)/I_\sigma$ and hence in $\rho(Z_\sigma)$. In fact, $\natural$ is surjective for if $z \in Z_\sigma$, then $\natural(L_z) = \rho(z)$ since

$$\langle [\varphi], \natural(L_z) \rangle = \langle [\varphi], L_z^*(1) \rangle$$

$$= \langle L_z[\varphi], \rho(e) \rangle$$

$$= \langle [\varphi], \rho(z) \times \rho(e) \rangle$$

$$= \langle [\varphi], \rho(z) \rangle$$

for $[\varphi] \in A(G)/I_\sigma$. Given that $\natural(\Lambda) = \natural(\Lambda')$, we have $(\Lambda'\Lambda^{-1})^*(1) = 1$ and as in the proof of [81, Proposition 5], one deduces that $\Lambda'\Lambda^{-1}$ is the identity map on $A(G)/I_\sigma$, so $\Lambda = \Lambda'$ and $\natural$ is injective (which gives $G_\sigma = \{L_z : z \in Z_\sigma\}$). It follows from [81, Proposition 4] that $\natural$ is a homeomorphism and condition (ii) in [81; p. 155] is satisfied.

(iii$'$) and (iv$'$). $A(G)/I_\sigma$ is a commutative Banach algebra such that $\{\varphi + I_\sigma : \varphi|_{Z_\sigma} \in C_c(Z_\sigma)\}$ is dense in it. Indeed, given $\varphi + I_\sigma \in A(G)/I_\sigma$, there is a sequence (φ_n) in $A(G) \cap C_c(G)$ norm-converging to φ. Then $\varphi_n|_{Z_\sigma} \in C_c(Z_\sigma)$ and $(\varphi_n + I_\sigma)$ converges to $\varphi + I_\sigma$ in $A(G)/I_\sigma$.

(v'). Since $I_\sigma^\perp$ is the w*-closed linear span of $\rho(Z_\sigma)$, Kaplansky's density theorem gives

$$\|\varphi + I_\sigma\| = \sup \left\{ \left| \sum_i \alpha_i \varphi(x_i) \right| : \alpha_i \in \mathbb{C}, \ x_i \in Z_\sigma \text{ and } \left\| \sum_i \alpha_i \rho(x_i) \right\| \leq 1 \right\}$$

for $\varphi + I_\sigma \in A(G)/I_\sigma$.

(vi'). Let $K \subset U \subset Z_\sigma$ where K is compact and U is open in Z_σ. Then there exists $\varphi + I_\sigma \in A(G)/I_\sigma$ such that $|\varphi| \leq 1$, $\varphi(x) = 1$ for $x \in K$ and $\varphi(x) = 0$ for $x \in Z_\sigma \backslash U$. This property of $A(G)$ is well-known [24; 3.2].

Now by [81; Theorem 6], $A(G)/I_\sigma$ is isometrically isomorphic to the Fourier algebra $A(\rho(Z_\sigma))$. $\qquad \square$

3.3. Jordan structures of harmonic functionals

We will now develop a structure theory for the space $I_\sigma^\perp$ of harmonic functionals for $\sigma \in B^1(G)$. Several results are non-commutative analogues of those for $J_\sigma^\perp$ with $\sigma \in M^1(G)$. In fact, $I_\sigma^\perp$ has non-associative Jordan structure whereas $J_\sigma^\perp$ is always (isometric to) a commutative von Neumann algebra. Jordan theory therefore plays a prominent role in our present development.

Our first task is to show that $I_\sigma^\perp$ is the range of a contractive projection on $VN(G)$ and hence admits a Jordan structure. By a *contractive projection* $P : VN(G) \to I_\sigma^\perp$ we mean a surjective linear map such that $P^2 = P$ and $\|P\| \leq 1$.

Proposition 3.3.1. *Let $\sigma \in B(G)$ and $\|\sigma\| = 1$. Then there is a contractive projection $P : VN(G) \to I_\sigma^\perp$ satisfying $\sigma \cdot P(T) = P(\sigma \cdot T)$ for every $T \in VN(G)$. If there is a w^*-w^*-continuous projection $Q : VN(G) \to I_\sigma^\perp$ satisfying $\sigma \cdot Q(T) = Q(\sigma \cdot T)$, then $P = Q$.*

Proof. For $n = 1, 2, \ldots$, define $\Lambda_n : VN(G) \to VN(G)$ by $\Lambda_n(T) = \sigma^n \cdot T$ for $T \in VN(G)$. Then Λ_n is w^*-w^*-continuous and contractive. Let $B(VN(G))$ be the locally convex space of bounded linear maps from $VN(G)$ to itself, equipped with the weak*-operator topology. Let $\mathcal{K}$ be the closed convex hull of $\{\Lambda_n : n = 1, 2, \ldots\}$ in $B(VN(G))$. Then $\mathcal{K}$ is compact. Define a map $\Phi : \mathcal{K} \to \mathcal{K}$ by $\Phi(\Lambda) = \sigma \cdot \Lambda$. Then Φ is affine and continuous. By Markov-Kakutani fixed-point theorem, there exists $P \in \mathcal{K}$ such that $\Phi(P) = P$. Then $P : VN(G) \to I_\sigma^\perp$ is the required projection. If $Q : VN(G) \to I_\sigma^\perp$ is a w^*-w^*-continuous projection and $\sigma \cdot Q(T) = Q(\sigma \cdot T)$ for $T \in VN(G)$, then $Q\Lambda = \Lambda Q$ for all $\Lambda \in \text{co}\{\Lambda_n : n = 1, 2, \ldots\}$ and hence $QP = PQ$. Therefore

$Q(T) = PQ(T) = QP(T) = P(T)$ for $T \in VN(G)$. $\qquad\qquad\square$

We refer to [31] for a result similar to Proposition 3.3.1. We will give some criteria for the above projection P to be w^*-w^*-continuous. By Remark 3.2.11, $I_\sigma^\perp$ is a von Neumann subalgebra of $VN(G)$ if $\sigma \in P^1(G)$. The following result may be of some independent interest.

Proposition 3.3.2. *Let* $\sigma \in P^1(G)$. *Then the projection* $P : VN(G) \to I_\sigma^\perp \subset VN(G)$ *in Proposition 3.3.1 is completely positive.*

Proof. We recall that a linear map Φ between C^*-algebras $\mathcal{A}$ and $\mathcal{B}$ is called *completely positive* if the maps $\Phi_n : M_n(\mathcal{A}) \to M_n(\mathcal{B})$ are all positive for $n = 1, 2, \ldots$, where $M_n(\mathcal{A})$ denotes the $n \times n$ matrix algebra over $\mathcal{A}$ and Φ_n is defined by $\Phi_n((a_{ij})) = (\Phi(a_{ij}))$. A $*$-homomorphism is completely positive and maps of the form $a \in \mathcal{A} \mapsto v^* a v \in \mathcal{B}$ are also completely positive for any bounded linear map v from a Hilbert space on which $\mathcal{B}$ acts to the one acted on by $\mathcal{A}$.

By [65; Theorem 6.4] and by the construction of P above, it suffices to show that the map $\Lambda : VN(G) \to VN(G)$ defined by

$$\Lambda(T) = \sigma \cdot T \quad (T \in VN(G))$$

is completely positive.

Let $\pi : G \to \mathcal{B}(H_\pi)$ be a cyclic representation such that $\sigma(x) = (\pi(x)\xi|\xi)$ for $x \in G$. Identify $L^2(G)$ as a closed subspace of $L^2(G) \otimes H_\pi$ via the embedding $h \in L^2(G) \overset{V}{\mapsto} h \otimes \xi \in L^2(G) \otimes H_\pi$. Let $f \in L^1(G)$ and $h, k \in L^2(G)$. We have

$$(\sigma \cdot \rho(g)h|k) = \int_G (\rho(x)h|k)\sigma(x)f(x)dx$$

$$= \int_G (\rho(x)h|k)\,(\pi(x)\xi|\xi)f(x)dx$$

$$= \int_G ((\rho \otimes \pi)(x)(h \otimes \xi)|k \otimes \xi)f(x)dx$$

$$= ((\rho \otimes \pi)(f)(h \otimes \xi)|k \otimes \xi)$$

$$= ((\rho \otimes \pi)(f)Vh|Vk)$$

$$= (V^*(\rho \otimes \pi)(f)Vh|k)$$

so $\Lambda(\rho(f)) = \sigma \cdot \rho(f) = V^*(\rho \otimes \pi)(f)V$.

We identify $L^1(G)$ with $\rho(L^1(G))$ in $VN(G)$. Since $\rho(L^1(G))$ is norm-dense in $C^*_r(G)$, the $*$-homomorphism $\rho \otimes \pi : \rho(L^1(G)) \to \mathcal{B}(L^2(G) \otimes H_\pi)$ extends to a $*$-homomorphism $\tau : C^*_r(G) \to \mathcal{B}(L^2(G) \otimes H_\pi)$ such that

$$\Lambda(T) = V^* \tau(T) V \quad (T \in C^*_r(G)).$$

Hence the restriction map

$$\Lambda^r = \Lambda \big|_{C^*_r(G)} : C^*_r(G) \to VN(G)$$

is completely positive. Since $C^*_r(G)$ is w^*-dense in $VN(G)$, the matrix C^*-algebra $M_n(C^*_r(G))$ is also w^*-dense in $M_n(VN(G))$ and by Kaplansky's density theorem, the positive cone $M_n(C^*_r(G))_+$ is w^*-dense in the cone $M_n(VN(G))_+$. Therefore the positivity of the map $\Lambda^r_n : M_n(C^*_r(G)) \to M_n(VN(G))$ implies the positivity of $\Lambda_n : M_n(VN(G)) \to M_n(VN(G))$ for $n = 1, 2, \ldots$, that is, Λ is completely positive. $\qquad\square$

Proposition 3.3.3. *Let* $\sigma \in B(G)$. *If there is a* w^*-w^*-*continuous bounded projection* $P : VN(G) \to I^\perp_\sigma$ *satisfying* $\varphi \cdot P(T) = P(\varphi \cdot T)$ *for* $\varphi \in A(G)$ *and* $T \in VN(G)$, *then the zero set* Z_σ *is open.*

Proof. Let $P : VN(G) \to I^\perp_\sigma$ be such a projection. Then it is the dual map of a continuous linear map $P_* : A(G)/I_\sigma \to A(G)$ between the preduals. For each $x \in Z_\sigma$, pick $\varphi_x \in A(G)$ such that $\varphi_x(x) \neq 0$. To show that Z_σ is open, we only need to establish

$$Z_\sigma = \bigcup_{x \in Z_\sigma} \{y \in G : P_*(\varphi_x + I_\sigma)(y) \neq 0\}.$$

First observe that $x \in Z_\sigma$ implies $\rho(x) \in I^\perp_\sigma$ and $P_*(\varphi_x + I_\sigma)(x) = \langle P_*(\varphi_x + I_\sigma), \rho(x) \rangle = \langle \varphi_x + I_\sigma, P(\rho(x)) \rangle = \langle \varphi_x + I_\sigma, \rho(x) \rangle = \varphi_x(x) \neq 0$.

Let $P_*(\varphi_x + I_\sigma)(y) \neq 0$. We show $y \in Z_\sigma$. Otherwise, $y \notin Z_\sigma$ and we can find $\psi \in I_\sigma$ such that $\psi(y) = 1$ by [38; 39.15]. It follows that $\psi \varphi_x + I_\sigma = I_\sigma$ and

$$0 = \langle P_*(\psi \varphi_x + I_\sigma), \rho(y) \rangle$$

$$= \langle \psi \varphi_x + I_\sigma, P(\rho(y)) \rangle$$

$$= \langle \varphi_x + I_\sigma, \psi \cdot P(\rho(y)) \rangle$$

$$= \langle \varphi_x + I_\sigma, P(\psi \cdot \rho(y)) \rangle$$

$$= \langle \varphi_x + I_\sigma, P(\rho(y)) \rangle \quad (\text{by } \psi(y) = 1)$$

$$= P_*(\varphi_x + I_\sigma)(y) \neq 0$$

which is impossible. So $y \in Z_\sigma$ and Z_σ is open. □

We have the following converse to Proposition 3.3.3 if the zero set Z_σ is a subgroup of G. The proof below also shows that the projection P in Proposition 3.3.1 has a simple form if Z_σ is an open subgroup of G.

Corollary 3.3.4. *Let $\sigma \in B(G)$ and let Z_σ be a subgroup of G. The following conditions are equivalent:*

(i) *Z_σ is open;*
(ii) *There is a w^*-w^*-continuous contractive projection $P : VN(G) \to I_\sigma^\perp$ such that $\varphi \cdot P(T) = P(\varphi \cdot T)$ for $\varphi \in A(G)$ and $T \in VN(G)$.*

Proof. We only need to prove (i) $\implies$ (ii). Since Z_σ is an open subgroup, the characteristic function χ_{Z_σ} belongs to $P(G)$. Define $P : VN(G) \to VN(G)$ by

$$P(T) = \chi_{Z_\sigma} \cdot T \quad (T \in VN(G)).$$

Write $\sigma_1 = \chi_{Z_\sigma}$. Then clearly P is a w^*-w^*-continuous contractive projection from $VN(G)$ onto $I_{\sigma_1}^\perp$ satisfying $\varphi \cdot P(T) = P(\varphi \cdot T)$. But $Z_{\sigma_1} = \sigma^{-1}\{1\} = Z_\sigma$ implies $I_{\sigma_1}^\perp = I_\sigma^\perp$ by Proposition 3.2.10. □

We have seen in Corollary 2.3.8 that the space of σ-harmonic functions $J_\sigma^\perp$ is isometric to an abelian von Neumann algebra via the contractive projection $P : L^\infty(G) \to J_\sigma^\perp$ on the abelian von Neumann algebra $L^\infty(G)$. In the non-commutative case of σ-harmonic functionals $I_\sigma^\perp$, the existence of a contractive projection on the generally non-commutative von Neumann algebra $VN(G)$ endows $I_\sigma^\perp$ with a non-associative algebraic structure, namely, the *Jordan triple structure* which involves a ternary product and is more elaborate than the associative binary product in $J_\sigma^\perp$. To exploit this ternary structure and to show some interesting connections to geometry, we give below a brief introduction to the theory of Jordan triples and background. We will only consider algebras over the complex field.

Jordan structures occur in symmetric Banach manifolds and operator algebras. A *Jordan algebra* is a commutative, but not necessarily associative, algebra whose elements satisfy the Jordan identity

$$a(ba^2) = (ab)a^2.$$

A *Jordan triple system* is a complex vector space Z with a *Jordan triple product*

$$\{\cdot, \cdot, \cdot\} : Z \times Z \times Z \to Z$$

which is symmetric and linear in the outer variables, conjugate linear in the middle variable and satisfies the Jordan triple identity

$$\{a, b, \{x, y, z\}\} = \{\{a, b, x\}, y, z\} - \{x, \{b, a, y\}, z\} + \{x, y, \{a, b, z\}\}.$$

A Jordan algebra with involution $*$ is a Jordan triple system with the usual Jordan triple product

$$\{a, b, c\} = (ab^*)c + (b^*c)a - (ca)b^*.$$

A complex Banach space Z is called a JB^*-*triple* if it is a Jordan triple system such that for each $z \in Z$, the linear map

$$z \,\square\, z : v \in Z \mapsto \{z, z, v\} \in Z$$

is Hermitian with non-negative spectrum and $\|z \,\square\, z\| = \|z\|^2$. A JB^*-triple Z is called a JBW^*-*triple* if it is a dual Banach space, in which case its predual is unique, denoted by Z_*, and the triple product is separately w^*-continuous. The second dual Z^{**} of a JB^*-triple is a JBW^*-triple.

The JB^*-triples form a large class of Banach spaces. They include for instance, C^*-algebras, Hilbert spaces and spaces of rectangular matrices. Indeed, a norm closed subspace Z of a C^*-algebra $\mathcal{A}$ is a JB^*-triple if $z \in Z$ implies $zz^*z \in Z$. In this case, the triple product is given by

$$\{x, y, z\} = \frac{1}{2} (xy^*z + zy^*x)$$

and Z is called a *subtriple* of $\mathcal{A}$. The fact that Z is closed with respect to the above triple product follows from the polarization formula:

$$16\{x, y, z\} = \sum_{\alpha^4 = 1 = \beta^2} \alpha\beta(x + \alpha y + \beta z)(x + \alpha y + \beta z)^*(x + \alpha y + \beta z). \tag{3.1}$$

JB^*-triples also include the class of JB^*-algebras. A JB^*-*algebra* is a complex Banach space which is also a Jordan algebra with involution $*$ satisfying $\|x^*\| = \|x\| = \|\{x, x^*, x\}\|^{1/3}$ where the triple product is given by

$$\{x, y, z\} = (xy^*)z + (zy^*)x - (xz)y^*.$$

Every C^*-algebra is a JB^*-algebra in the following Jordan product:

$$a \circ b = \frac{1}{2}(ab + ba).$$

A JB^*-subalgebra of a C^*-algebra, with respect to the product $\circ$ above, is called a JC^*-*algebra*. A JB^*-algebra is called a JBW^*-*algebra* if it is a dual Banach space. JBW^*-algebras which are isometric to JC^*-algebras are called JW^*-*algebras*.

Given a JB^*-triple Z, an element $e \in Z$ is called a *tripotent* if $\{e, e, e\} = e$. A tripotent $e \in Z$ is called *unitary* if $\{eez\} = z$ for all $z \in Z$. Tripotents in C^*-algebras are exactly the partial isometries. If a JB^*-triple Z contains a unitary tripotent u, then it is a JB^*-algebra in the following Jordan product and involution:

$$x \circ y = \{x, u, y\}, \quad x^* = \{u, x, u\}$$

with u as an identity. A JBW^*-triple is the norm-closed linear span of its tripotents.

In geometry, JB^*-triples arise as tangent spaces to complex symmetric Banach manifolds which are infinite-dimensional generalization of the Hermitian symmetric spaces classified by E. Cartan [10] in the 1930s using Lie groups. An (*analytic*) *Banach manifold* is a manifold modelled locally on open subsets of Banach spaces such that the coordinate transformations are bianalytic maps. The manifold is called *real* (*complex*) if the underlying Banach spaces are real (complex). A Banach manifold M is *symmetric* if every point $a \in M$ is an isolated fixed-point of a symmetry $s_a : M \to M$ which is a bianalytic map satisfying $s_a^2 = Id$. Given a connected complex symmetric manifold M and $a \in M$, Kaup [48] has shown that the symmetry s_a induces a Jordan triple structure on the tangent space $T_a M$ at a. Further, if M is a *bounded domain*, that is, a bounded open connected set in a complex Banach space, then $T_a M$ can be given a norm such that $T_a M$ is a JB^*-triple and M is biholomorphically equivalent to the open unit ball of $T_a M$, as shown in [48]. Conversely, the open unit ball of every JB^*-triple is a complex symmetric Banach manifold. Here the crucial fact is that the geometry of a JB^*-triple is completely determined by its triple product in that the surjective linear isometries between JB^*-triples are exactly the *triple isomorphisms* which are bijective linear maps preserving the triple product [48,42]. Also, two JB^*-triples are isometric if, and only if, their open unit balls are biholomorphically equivalent. We refer to [13,72,80] for further references concerning JB^*-triples and symmetric Banach manifolds.

JB^*-triples also appear as the ranges of contractive projections on C^*-algebras as shown in [25]. Using this result and Proposition 3.3.1, we show below that $I_\sigma^\perp$ is a JW^*-algebra and therefore its open unit ball is a complex symmetric manifold. We denote by $\mathcal{B}(H)$ the von Neumann algebra of bounded linear operators on a Hilbert space H.

Proposition 3.3.5. *Let* $\sigma \in B(G)$ *and* $\|\sigma\| = 1$. *Then* $I_\sigma^\perp$ *is a* JW^*-*algebra.*

Proof. Let $P : VN(G) \to I_\sigma^\perp$ be the contractive projection in Proposition 3.3.1. By [25], $I_\sigma^\perp$ is a JB^*-triple with the triple product

$$\{u, v, w\} = \frac{1}{2} P(uv^*w + wv^*u)$$

for $u, v, w \in I_\sigma^\perp$, and moreover, $I_\sigma^\perp$ is isometric to a subtriple of a C^*-algebra. Since $I_\sigma^\perp$ is a dual Banach space, it follows from [7; Corollary 9] that $I_\sigma^\perp$ is isometric to a w^*-closed subtriple of $\mathcal{B}(H)$ for some Hilbert space H. Pick any $x \in Z_\sigma$. Then $u = \rho(x)$ is a unitary element in $VN(G)$. Therefore $w \in I_\sigma^\perp$ implies $\{u, u, w\} = P(w) = w$ and by the above remarks, $I_\sigma^\perp$ is a JW^*-algebra in the following Jordan product and involution:

$$v \circ w = \{v, u, w\} = \frac{1}{2} P(vu^*w + wu^*v)$$

$$v^\# = \{u, v, u\} = P(uv^*u).$$

□

We will now study the Jordan structures of $I_\sigma^\perp$ in detail. We note that, as the norm of $I_\sigma^\perp$ is always fixed, the triple structure of $I_\sigma^\perp$ is unique by the previous remarks.

Definition 3.3.6. Given $\sigma \in B(G)$ with $\|\sigma\| = 1$, we denote by $P_\sigma : VN(G) \to I_\sigma^\perp$ the contractive projection constructed in Proposition 3.3.1 and if $I_\sigma^\perp \neq \{0\}$, we always fix a unitary element $u \in I_\sigma^\perp$ so that $(I_\sigma^\perp, u)$ is a JW^*-algebra with identity u, in the Jordan product $\circ$ and involution $\#$ defined in Proposition 3.3.5.

We note that $I_\sigma^\perp$ need not be a subtriple of $VN(G)$ in general and we will clarify this situation later. We first look at the case of $I_\sigma^\perp$ being a subalgebra of $VN(G)$ more closely.

Lemma 3.3.7. *Let $\sigma \in B(G)$ with $\|\sigma\| = 1$ and the induced projection $P_\sigma :$ $VN(G) \to I_\sigma^\perp$. The following conditions are equivalent:*

(i) $P_\sigma(1) = 1$ *where 1 is the identity of $VN(G)$;*

(ii) $P_\sigma(1) \neq 0$;

(iii) $\sigma(e) = 1$;

(iv) $I_\sigma^\perp$ *is a nontrivial subalgebra of $VN(G)$;*

(v) Z_σ *is a subgroup of G.*

Proof. (i) $\Longleftrightarrow$ (ii). Since $P_\sigma(1)$ is in the w^*-closed convex hull of $\{\sigma^n \cdot 1 : n \in \mathbb{N}\}$ and $\sigma^n \cdot 1 = \sigma(e)^n 1$, we have $P_\sigma(1) = \alpha 1$ for some $\alpha \in \mathbb{C}$. Therefore $P_\sigma(1) \neq 0$ if, and only if, $1 \in I_\sigma^\perp$, that is, $P_\sigma(1) = 1$.

(i) $\Longleftrightarrow$ (iii). From above, $\sigma(e) = 1$ implies that $P_\sigma(1) = 1$. The converse follows from $1 = P_\sigma(1) = \sigma \cdot P_\sigma(1) = \sigma \cdot 1 = \sigma(e)1$.

By Proposition 3.2.10 and Remark 3.2.11, it remains to show (iv) $\Longrightarrow$ (ii). Suppose $P_\sigma(1) = 0$. Since $I_\sigma^\perp \neq \{0\}$, there exists a unitary $u \in I_\sigma^\perp$. Then $u^2 \in I_\sigma^\perp$ and by the property of the contractive projection P_σ (cf.[72,p.229]), we have

$$u^2 = P_\sigma(u^2) = P_\sigma(u1u) = P_\sigma(uP_\sigma(1)^*u) = 0$$

which is impossible. $\qquad\qquad\qquad\qquad\qquad\qquad\qquad\qquad\qquad\qquad\Box$

Remark 3.3.8. In the proof of (iv) $\Longrightarrow$ (ii) above, we did not assume that $I_\sigma^\perp$ is a $*$-subalgebra of $VN(G)$ with identity 1 although the result implies this. Given that $I_\sigma^\perp$ is a $*$-subalgebra of $VN(G)$, we would have $P_\sigma(vTw) = vP_\sigma(T)w$ for $v, w \in I_\sigma^\perp$ and $T \in VN(G)$, by a well-known result of Tomiyama [77]. We note that the condition $\|\sigma\| = 1$ in Lemma 3.3.7 is not assumed in Theorem 3.2.12.

Proposition 3.3.9. *Let $\sigma \in B(G)$ and $\|\sigma\| = 1$. If $I_\sigma^\perp$ is a subtriple of $VN(G)$, then the zero set $Z_\sigma = \sigma^{-1}\{1\}$ satisfies the following condition:*

$$x, y, z \in Z_\sigma \quad \text{implies} \quad xy^{-1}z \in Z_\sigma.$$

The converse holds if $I_\sigma^\perp$ is the w^*-closed linear span of $\rho(Z_\sigma)$.

Proof. Let $I_\sigma^\perp$ be a subtriple of $VN(G)$ and let $x_1, x_2, x_3 \in Z_\sigma$. We show $x_1 x_2^{-1} x_3 \in Z_\sigma$. Suppose not, we deduce a contradiction. Let $T = \rho(x_1) + \rho(x_2) + \rho(x_3) \in I_\sigma^\perp$. Since $I_\sigma^\perp$ is a subtriple of $VN(G)$, we have

$$\sum_{i,j,k} \rho(x_i x_j^{-1} x_k) = TT^*T \in I_\sigma^\perp.$$

By [38; 39.15], we can choose $\varphi \in I_\sigma$ such that $\varphi(x_i x_j^{-1} x_k) = 1$ if $x_i x_j^{-1} x_k \notin Z_\sigma$. It follows that

$$0 = \left\langle \varphi, \sum_{i,j,k} \rho(x_i x_j^{-1} x_k) \right\rangle = \sum_{x_i x_j^{-1} x_k \notin Z_\sigma} \varphi(x_i x_j^{-1} x_k) \neq 0$$

which is a contradiction. Hence $x_1 x_2^{-1} x_3 \in Z_\sigma$.

Conversely, suppose $I_\sigma^\perp = \overline{\text{span}}\, \rho(Z_\sigma)$ is the w^*-closed linear span of $\rho(Z_\sigma)$. Recall that the triple product of $VN(G)$ is defined by

$$\{R, S, T\} = \frac{1}{2}(RS^*T + TS^*R).$$

Given $T \in I_\sigma^\perp$, we need to show $TT^*T \in I_\sigma^\perp$. If $T = \sum_i \alpha_i \rho(x_i) \in \text{span}\, \rho(Z_\sigma)$, then

$$TT^*T = \sum_{i,j,k} \alpha_i \overline{\alpha_j} \alpha_k \rho(x_i x_j^{-1} x_k) \in \text{span}\, \rho(Z_\sigma)$$

since Z_σ satisfies the given condition. So the norm closure of $\text{span}\, \rho(Z_\sigma)$ is a subtriple of $VN(G)$ and standard arguments imply that the w^*-closure $\overline{\text{span}}\, \rho(Z_\sigma)$ is also a subtriple of $VN(G)$. We give below these arguments for completeness.

First note that for $R, S, T \in \text{span}\, \rho(Z_\sigma)$, the polarization formula (3.1) implies $\{R, S, T\} \in \text{span}\, \rho(Z_\sigma)$. Now let $T \in \overline{\text{span}}\, \rho(Z_\sigma)$ with $T = w^* - \lim_\alpha T_\alpha$ and $T_\alpha \in \text{span}\, \rho(Z_\sigma)$. Then $T^* = w^* - \lim_\alpha T_\alpha^*$ and $\{T_\gamma, T, T_\beta\} = \frac{1}{2}(T_\gamma T^* T_\beta + T_\beta T^* T_\gamma) = w^* - \lim_\alpha \frac{1}{2}(T_\gamma T_\alpha^* T_\beta + T_\beta T_\alpha^* T_\gamma) \in \overline{\text{span}}\, \rho(Z_\sigma)$ for fixed β, γ. Next, $\{T_\gamma, T, T\} = w^* - \lim_\beta \{T_\gamma, T, T_\beta\} \in \overline{\text{span}}\, \rho(Z_\sigma)$ for fixed γ. Finally $\{T, T, T\} = w^* - \lim_\gamma \{T_\gamma, T, T\} \in \overline{\text{span}}\, \rho(Z_\sigma)$. $\qquad\square$

Remark 3.3.10. We have seen that if Z_σ is a subgroup of G, then $I_\sigma^\perp$ is the w^*-closed linear span of $\rho(Z_\sigma)$. The latter is also true if Z_σ is a spectral set for $A(G)$, for instance, a finite set. We recall that a closed subset $E \subset G$ is called a *spectral set* for $A(G)$ if the ideal $\{\varphi \in A(G) : \varphi(E) = \{0\}\}$ is the only closed ideal whose zero set is E.

We have seen in Chapter 2 that sharper results for harmonic functions can be obtained by studying those functions which are uniformly continuous. Now we investigate the non-commutative analogue of these functions. They are the harmonic functionals belonging to $\overline{\operatorname{span}}\, A(G) \cdot VN(G)$ as the latter identifies with the bounded uniformly continuous functions on $\widehat{G}$ when G is abelian (cf. [32,33]). Results for $J_\sigma^\perp \cap C_{\ell u}(G)$ have non-commutative analogues for $I_\sigma^\perp \cap \overline{\operatorname{span}}\, A(G) \cdot VN(G)$.

Analogous to (2.2), given $m \in \big(\overline{\operatorname{span}}\, A(G) \cdot VN(G)\big)^*$ and $T \in \overline{\operatorname{span}}\, A(G) \cdot VN(G)$, we define $m \circ T \in A(G)^* = VN(G)$ by

$$\langle \varphi, m \circ T \rangle = \langle \varphi \cdot T, m \rangle \quad (\varphi \in A(G)).$$

Then we have $m \circ T \in \overline{\operatorname{span}}\, A(G) \cdot VN(G)$ which also contains the w^*-closed convex hull K_T of the set $\{\varphi \cdot T : \varphi \in A(G), \|\varphi\| \leq 1\}$ (see [53, Lemma 5.1]).

By [56], there is a linear (into) isometry $\eta \in B_\rho(G) \mapsto \widetilde{\eta} \in \big(\overline{\operatorname{span}}\, A(G) \cdot VN(G)\big)^*$ such that $\widetilde{\eta}(\varphi \cdot T) = \langle \varphi, \eta \cdot T \rangle$ for $\varphi \in A(G)$ and $T \in VN(G)$. Moreover, we have

$$\langle m \circ T, \widetilde{\eta} \rangle = \langle \widetilde{\eta} \circ T, m \rangle \tag{3.2}$$

for $m \in \big(\overline{\operatorname{span}}\, A(G) \cdot VN(G)\big)^*$ and $\eta \in B_\rho(G)$.

Let $P_\rho^1(G) = \{\eta \in P_\rho(G) : \eta(e) = 1\} \subset B_\rho(G) \subset B(G)$ and let $P_\rho^1(G)$ inherit the w^*-topology $\sigma\big(B(G), C^*(G)\big)$ which coincides with the topology $\sigma\big(P_\rho^1(G), L^1(G)\big)$ since $P_\rho^1(G)$ is norm-bounded and $L^1(G)$ is norm-dense in $C^*(G)$. Let (η_α) be a net in $P_\rho^1(G)$ w^*-converging to $\eta \in P_\rho^1(G)$. Then by [60,34], we have $\|\varphi\eta_\alpha - \varphi\eta\| \to 0$ for all $\varphi \in A(G)$ which implies that $\widetilde{\eta}_\alpha(\varphi \cdot T) = \langle \varphi, \eta_\alpha \cdot T \rangle \to \langle \varphi, \eta \cdot T \rangle = \widetilde{\eta}(\varphi \cdot T)$ for $\varphi \in A(G)$ and $T \in VN(G)$. So $\widetilde{\eta}_\alpha \to \widetilde{\eta}$ in the w^*-topology of $\big(\overline{\operatorname{span}}\, A(G) \cdot VN(G)\big)^*$. Conversely, if $\widetilde{\eta}_\alpha \to \widetilde{\eta}$ in the w^*-topology, then $\eta_\alpha \to \eta$ in the topology $\sigma\big(P_\rho^1(G), L^1(G)\big)$. We write $\widetilde{P}_\rho^1(G) = \{\widetilde{\eta} : \eta \in P_\rho^1(G)\}$.

Lemma 3.3.11. *Let* $T \in \operatorname{span} A(G) \cdot VN(G)$ *and let* K_T *be the* w^*-*closed convex hull of* $\{\varphi \cdot T : \varphi \in A(G), \|\varphi\| \leq 1\}$. *Then* K_T *is equicontinuous on* $\widetilde{P}_\rho^1(G)$.

Proof. Let $(\widetilde{\eta}_\alpha)$ be a net in $\widetilde{P}_\rho^1(G)$ w^*-converging to $\widetilde{\eta} \in \widetilde{P}_\rho^1(G)$. Let $\varepsilon > 0$. We show

$$|\langle s, \widetilde{\eta}_\alpha \rangle - \langle s, \widetilde{\eta} \rangle| < \varepsilon \tag{3.3}$$

for all $S \in K_T$, from some α onwards. It can be seen from the following arguments that we may assume $T \in A(G) \cdot VN(G)$ with $T = \varphi' \cdot T'$ where $\varphi' \in A(G)$ and $T' \in VN(G)$. By w^*-convergence of (η_α) and the above remark, we have $\|\varphi' \eta_\alpha - \varphi' \eta\| \to 0$.

Let $S \in K_T$. Then there is a net (φ_β) in $A(G)$ with $\|\varphi_\beta\| \leq 1$ such that $(\varphi_\beta \cdot T)$ w^*-converges to S. Now $(\widetilde{\varphi}_\beta)$ is a net in $\left(\overline{\text{span}}\, A(G) \cdot VN(G)\right)^*$ and we may suppose, passing to a subnet if necessary, that $(\widetilde{\varphi}_\beta)$ w^*-converges to $m \in \left(\overline{\text{span}}\, A(G) \cdot VN(G)\right)^*$ say. For $\varphi \in A(G)$, we have $\langle \varphi, m \circ T \rangle = \langle \varphi \cdot T, m \rangle = \lim_\beta \langle \varphi \cdot T, \widetilde{\varphi}_\beta \rangle = \lim_\beta \langle \varphi_\beta, \varphi \cdot T \rangle = \lim_\beta \langle \varphi, \varphi_\beta \cdot T \rangle = \langle \varphi, S \rangle$, that is, $S = m \circ T$. It follows that

$$
\begin{aligned}
|\langle S, \widetilde{\eta}_\alpha \rangle - \langle S, \widetilde{\eta} \rangle| &= |\langle m \circ T, \widetilde{\eta}_\alpha - \widetilde{\eta} \rangle| \\
&= |\langle (\widetilde{\eta}_\alpha - \widetilde{\eta}) \circ T, m \rangle| \quad \text{by (3.2)} \\
&= \lim_\beta |\langle (\widetilde{\eta}_\alpha - \widetilde{\eta}) \circ T, \widetilde{\varphi}_\beta \rangle| \\
&= \lim_\beta |\langle \varphi_\beta, (\widetilde{\eta}_\alpha - \widetilde{\eta}) \circ T \rangle| \\
&= \lim_\beta |\langle \varphi_\beta \cdot T, \widetilde{\eta}_\alpha - \widetilde{\eta} \rangle| \\
&= \lim_\beta |\langle \eta_\alpha - \eta, \varphi_\beta \cdot T \rangle| \\
&= \lim_\beta |\langle \varphi' \eta_\alpha - \varphi' \eta, \varphi_\beta \cdot T' \rangle| \\
&\leq \|T'\| \, \|\varphi' \eta_\alpha - \varphi' \eta\| \to 0
\end{aligned}
$$

which clearly yields (3.3). $\qquad\qquad\qquad\qquad\qquad\qquad\qquad\qquad\qquad\square$

We now prove a non-associative analogue of Theorem 2.2.17.

Theorem 3.3.12. *Let $\sigma \in B_\rho(G)$ and $\|\sigma\| = 1$. Then there exists a net (μ_α) in the convex hull of $\{\sigma^n : n \in \mathbb{N}\}$ such that for $R, S, T \in I_\sigma^\perp \cap VN_c(G)$, their Jordan triple product is given by*

$$
2\{R, S, T\} = \lim_\alpha \mu_\alpha \cdot (RS^*T + TS^*R)
$$

uniformly on compact subsets of $\widetilde{P}_\rho^1(G)$. Further, if G is amenable, then the above convergence holds for all $R, S, T \in I_\sigma^\perp \cap \overline{\text{span}}\, A(G) \cdot VN(G)$.

Proof. By the construction of the projection $P_\sigma : VN(G) \to I_\sigma^\perp$ in Proposition 3.3.1, there is a net (μ_α) in the convex hull co $\{\sigma^n : n \geq 1\}$ such that

$$\langle \varphi, \{R, S, T\} \rangle = \lim_\alpha \left\langle \varphi, \mu_\alpha \cdot \left(\frac{1}{2} (RS^*T + TS^*R) \right) \right\rangle$$

for $\varphi \in A(G)$ and $R, S, T \in I_\sigma^\perp$.

Let $V = \frac{1}{2}(RS^*T + TS^*R)$ where $R, S, T \in I_\sigma^\perp \cap VN_c(G)$. Then $V = \varphi \cdot V$ whenever $\varphi \in A(G)$ and $\varphi = 1$ on supp V. Let K_V be the w^*-closed convex hull of $\{\varphi \cdot V : \varphi \in A(G), \|\varphi\| \leq 1\}$. Then K_V is equicontinuous on $\widetilde{P}_\rho^1(G)$ by Lemma 3.3.11.

We first show that $\{R, S, T\} \in K_V$. For this, it suffices to show $\mu_\alpha \cdot V \in K_V$ for all α. We note that $\|\mu_\alpha\| \leq 1$ and $\widetilde{\mu}_\alpha \in (\overline{\text{span}} \, A(G) \cdot VN(G))^*$ has a norm-preserving extension $\widetilde{\widetilde{\mu}}_\alpha \in VN(G)^*$. Regarding $A(G) \subset VN(G)^*$ and by Goldstein's theorem, there is a net (φ_β) in $A(G)$, with $\|\varphi_\beta\| \leq 1$, w^*-converging to $\widetilde{\widetilde{\mu}}_\alpha$. Then $(\varphi_\beta \cdot V)$ w^*-converges to $(\mu_\alpha \cdot V)$ in $VN(G)$, that is, $\mu_\alpha \cdot V \in K_V$.

We now show that $(\mu_\alpha \cdot V)$ converges pointwise to $\{R, S, T\}$ on $\widetilde{P}_\rho^1(G)$. Let $\widetilde{\nu} \in \widetilde{P}_\rho^1(G)$. Let $\varepsilon > 0$. By equicontinuity of K_V, there is a w^*-neighbourhood N of $\widetilde{\nu}$ such that

$$|\langle W, \widetilde{\varphi} \rangle - \langle W, \widetilde{\nu} \rangle| < \varepsilon$$

for all $\widetilde{\varphi} \in N$ and $W \in K_V$. Choosing a net in $A(G)$ w^*-converging to $\widetilde{\widetilde{\nu}}$ as above, we can pick $\varphi \in A(G)$ such that $\widetilde{\varphi} \in N$. There exists α_0 such that $\alpha \geq \alpha_0$ implies

$$|\langle \varphi, \{R, S, T\} \rangle - \langle \varphi, \mu_\alpha \cdot V \rangle| < \varepsilon$$

and therefore

$$|\langle \{R, S, T\}, \widetilde{\nu} \rangle - \langle \mu_\alpha \cdot V, \widetilde{\nu} \rangle| \leq |\langle \{R, S, T\}, \widetilde{\nu} \rangle - \langle \{R, S, T\}, \widetilde{\varphi} \rangle|$$
$$+ |\langle \{R, S, T\}, \widetilde{\varphi} \rangle - \langle \mu_\alpha \cdot V, \widetilde{\varphi} \rangle|$$
$$+ |\langle \mu_\alpha \cdot V, \widetilde{\varphi} \rangle - \langle \mu_\alpha \cdot V, \widetilde{\nu} \rangle| < 3\varepsilon.$$

This proves $\mu_\alpha \cdot V \to \{R, S, T\}$ pointwise on $\widetilde{P}_\rho^1(G)$. Since $\widetilde{P}_\rho^1(G)$ is locally compact and $|\langle W, \widetilde{\nu} \rangle| \leq \|V\|$ for all $W \in K_V$ and $\widetilde{\nu} \in \widetilde{P}_\rho^1(G)$, the convergence is uniform on compact subsets of $\widetilde{P}_\rho^1(G)$ by equicontinuity of K_V (cf. [49; p. 232]).

Finally, if G is amenable, we have $\overline{\text{span}}\,A(G) \cdot VN(G) = A(G) \cdot VN(G)$ as remarked before and hence the above arguments apply to all $R, S, T \in I_\sigma^\perp \cap \overline{\text{span}}\,A(G) \cdot VN(G)$.

$\square$

For amenable groups G, we have the following description of the "*uniformly continuous*" harmonic functionals.

Proposition 3.3.13. *Let G be amenable and let $\sigma \in B(G)$. Then*

$$I_\sigma^\perp \cap \left(A(G) \cdot VN(G) \right) = \overline{I_\sigma^\perp \cap VN_c(G)}$$

$$= \{ \varphi \cdot T : \varphi \in A(G), T \in I_\sigma^\perp \}.$$

Proof. We first note that $\overline{\text{span}}\,A(G) \cdot I_\sigma^\perp = \overline{I_\sigma^\perp \cap VN_c(G)}$. Indeed, if $T \in I_\sigma^\perp \cap VN_c(G)$, then $T = \varphi \cdot T \in A(G) \cdot I_\sigma^\perp$ where $\varphi \in A(G)$ and $\varphi = 1$ on supp T. Conversely, given $\psi \cdot S \in A(G) \cdot I_\sigma^\perp$, choose a sequence (ψ_n) in $A(G) \cap C_c(G)$ with $\|\psi_n - \psi\| \to 0$. Then $\|\psi_n \cdot S - \psi \cdot S\| \to 0$ and $\psi_n \cdot S \in I_\sigma^\perp \cap VN_c(G)$.

To complete the proof, we need to show $I_\sigma^\perp \cap \left(A(G) \cdot VN(G) \right) \subset A(G) \cdot I_\sigma^\perp$. Since G is amenable, the existence of a bounded approximate identity (u_α) in $A(G)$ and Cohen's factorization imply that $A(G) \cdot I_\sigma^\perp = \overline{\text{span}}\,A(G) \cdot I_\sigma^\perp$ [38; 32.22]. Hence $T \in I_\sigma^\perp \cap \left(A(G) \cdot VN(G) \right)$, with $T = \varphi \cdot S$ where $\varphi \in A(G)$ and $S \in VN(G)$, implies that $u_\alpha \cdot (\varphi \cdot S) = u_\alpha \cdot T \in A(G) \cdot I_\sigma^\perp$ and $\varphi \cdot S = \lim_\alpha u_\alpha \cdot (\varphi \cdot S) \in \overline{A(G) \cdot I_\sigma^\perp} = A(G) \cdot I_\sigma^\perp$. $\square$

Proposition 3.3.14. *Let $\sigma \in B(G)$ and let Z_σ be a subgroup of G. Then $I_\sigma^\perp \cap \overline{\text{span}}\,A(G) \cdot VN(G) = \overline{I_\sigma^\perp \cap VN_c(G)}$.*

Proof. By Proposition 3.2.10, $A(G)/I_\sigma$ identifies with the Fourier algebra $A(Z_\sigma)$. Let $\Phi : VN(Z_\sigma) \to I_\sigma^\perp$ be the induced isometric isomorphism. By [47; Lemma 3.2], we have $\Phi\left(\overline{\text{span}}\,A(Z_\sigma) \cdot VN(Z_\sigma) \right) = I_\sigma^\perp \cap \overline{\text{span}}\,A(G) \cdot VN(G)$ and $\Phi(VN_c(Z_\sigma)) = I_\sigma^\perp \cap VN_c(G)$. Since $\overline{\text{span}}\,A(Z_\sigma) \cdot VN(Z_\sigma) = \overline{VN_c(Z_\sigma)}$, we have $I_\sigma^\perp \cap \overline{\text{span}}\,A(G) \cdot VN(G) = \overline{I_\sigma^\perp \cap VN_c(G)}$. $\square$

Now we discuss some other algebraic and geometric consequences of the Jordan structures of $I_\sigma^\perp$.

Given $\sigma \in B(G)$ and $\|\sigma\| = 1$, we embed the Banach algebra $A(G)/I_\sigma$ in its second dual $\left(A(G)/I_\sigma \right)^{**}$ in the usual way, where the latter is equipped with the first Arens product

$$\xi \eta = w^* - \lim_\alpha \lim_\beta \xi_\alpha \eta_\beta$$

where $\xi, \eta \in \left(A(G)/I_\sigma\right)^{**}$ are w^*-limits of the nets (ξ_α) and (η_β) respectively in $A(G)/I_\sigma$. The product $\xi\eta$ is w^*-continuous in ξ for fixed η. Then the *algebraic centre* $\mathcal{Z}$ of $\left(A(G)/I_\sigma\right)^{**}$ coincides with the set $\{\xi \in \left(A(G)/I_\sigma\right)^{**} : \eta \mapsto \xi\eta$ is w^*-continuous$\}$. $A(G)/I_\sigma$ is *Arens regular* if $\mathcal{Z} = \left(A(G)/I_\sigma\right)^{**}$.

Proposition 3.3.15. *Let $\sigma \in B(G)$ and $\|\sigma\| = 1$. Given that $A(G)/I_\sigma$ has bounded approximate identity and Z_σ is discrete, we have*

   (i) $A(G)/I_\sigma$ *coincides with the centre of* $\left(A(G)/I_\sigma\right)^{**}$;
   (ii) $A(G)/I_\sigma$ *is Arens regular if, and only if, $I_\sigma^\perp$ is linearly isomorphic to a Hilbert space.*

Proof. (i) Since $\left(A(G)/I_\sigma\right)^* = I_\sigma^\perp$ is a JW^*-algebra, its predual $A(G)/I_\sigma$ is isometrically isomorphic to a complemented subspace of the predual of a von Neumann algebra, by [15; Theorem 2], and is therefore weakly sequentially complete (cf. [74; Corollary 5.2]). Since Z_σ is discrete, $A(G)/I_\sigma$ is an ideal in $\left(A(G)/I_\sigma\right)^{**}$ [30, Theorem 3]. Hence $A(G)/I_\sigma = \mathcal{Z}$ by [6; Theorem 2.1(iii)].

   (ii) Arens regularity of $A(G)/I_\sigma$ is equivalent to reflexivity by (i). It is well-known that a JB^*-triple is reflexive if, and only if, it is linearly isomorphic to a Hilbert space (cf.[19]). $\square$

Remark 3.3.16. It is unknown if (ii) above remains valid without the existence of a bounded approximate identity in $A(G)/I_\sigma$, even in the special case of $I_\sigma = \{0\}$ where $\sigma^{-1}\{1\} = G$. Also we do not know if Arens regularity of $A(G)/I_\sigma$ would actually imply that Z_σ is finite in Proposition 3.3.15(ii). However, as shown in [31], if G is a free group on two generators, then there is a closed ideal I in $A(G)$ which generalizes I_σ such that $A(G)/I$ is linearly isomorphic to an infinite-dimensional separable Hilbert space and hence Arens regular.

Finally in this section we study the geometry of $I_\sigma^\perp$. We have already noted that the open unit ball U_σ of $I_\sigma^\perp$ is a symmetric Banach manifold. The classification of U_σ, or rather, of $I_\sigma^\perp$, will be carried out in the next section. We study here the geometry of certain components of the boundary of U_σ which may be useful for the understanding of the boundary structure of U_σ. We first consider the finite-dimensional case. The simplest example is that dim $I_\sigma^\perp = 2$ in which case U_σ is a bidisc in $\mathbb{C}^2$ as will be seen presently.

By the structure theory of Jordan algebras (cf. [43]), if dim $I_\sigma^\perp < \infty$, then it is (isomorphic to) a finite ℓ_∞-sum of JW^*-algebras, each of which is one of the following Cartan factors:

   (i) M_n ($n \times n$ complex matrices)

   (ii) A_{2n} ($2n \times 2n$ antisymmetric complex matrices, $n \geq 2$)

   (iii) S_n ($n \times n$ symmetric complex matrices, $n \geq 2$)

   (iv) V (spin factor)

where V is a subspace of M_n, of dimension ≥ 3, such that $a \in V$ implies that $a^* \in V$ and a^2 is a scalar multiple of the identity matrix. Now if $\dim I_\sigma^\perp = 2$, then $I_\sigma^\perp = \mathbb{C} \oplus_{\ell_\infty} \mathbb{C}$ since $\dim A_{2n}, \dim S_n, \dim V > 2$. So $U_\sigma = \{(\alpha, \beta) \in \mathbb{C}^2 : |\alpha|, |\beta| < 1\}$. If we take a discrete group G with an element x satisfying $x^2 = e$ and if we let $\sigma = \chi_{\{x,e\}} \in B(G)$, then $\|\sigma\| = 1$ and $I_\sigma^\perp = \{\alpha \rho(x) + \beta 1 : \alpha, \beta \in \mathbb{C}\}$ which is 2-dimensional. One can find simple examples of a non-abelian $I_\sigma^\perp$. Indeed, if $\sigma^{-1}\{1\}$ is a subgroup, of order 6, of G, then $I_\sigma^\perp$ is a von Neumann subalgebra of $VN(G)$ and $\dim I_\sigma^\perp = 6$. The above classification list implies that $I_\sigma^\perp$ is either $\mathbb{C}^6$ or $\mathbb{C}^2 \oplus M_2$, and the latter occurs if $\sigma^{-1}\{1\}$ is non-abelian. It may be interesting to note that dimension analysis also leads to the following result.

Lemma 3.3.17. *Given $x \in G$ with $x^2 \neq e$, then there does not exist $\sigma \in B(G)$ satisfying both $\|\sigma\| = 1$ and $\sigma^{-1}\{1\} = \{x, x^2\}$.*

Proof. Suppose there exists $\sigma \in B(G)$ satisfying the given conditions, then $I_\sigma^\perp = \text{span}\{\rho(x), \rho(x^2)\}$ is an abelian von Neumann algebra. We deduce a contradiction by showing that $I_\sigma^\perp$ is a Cartan factor. For this, we show that $I_\sigma^\perp$ does not contain nontrivial triple ideal. A (closed) subspace $J \subset I_\sigma^\perp$ is a *triple ideal* if $\{I_\sigma^\perp, I_\sigma^\perp, J\} \subset J$. Since $\|\sigma\| = 1$, there is a contractive projection $P_\sigma : VN(G) \to I_\sigma^\perp$.

    We show that no 1-dimensional subspace of $I_\sigma^\perp$ is an ideal. First, $\mathbb{C}\rho(x)$ is not an ideal because $\{\rho(x^2), \rho(x), \rho(x)\} = \rho(x^2) \notin \mathbb{C}\rho(x)$. Likewise $\mathbb{C}\rho(x^2)$ is not an ideal. Next let $u = \alpha\rho(x) + \beta\rho(x^2)$ and $J = \mathbb{C}u$, where $\alpha, \beta \in \mathbb{C}\backslash\{0\}$. Then $\{\rho(x), \rho(x^2), u\} = \alpha\{\rho(x), \rho(x^2), \rho(x)\} + \beta\{\rho(x), \rho(x^2), \rho(x^2)\} = \alpha P_\sigma(\rho(x)\rho(x^2)^*\rho(x)) + \beta\rho(x) = \alpha P_\sigma(1) + \beta\rho(x) = \beta\rho(x) \notin J$, by Proposition 3.3.7. So J is not an ideal either. $\quad\square$

    As $\dim I_\sigma^\perp$ increases, it can be seen that the geometric complexity of U_σ increases rapidly. We will study the connected components of projections in $I_\sigma^\perp$. These are symmetric Banach manifolds contained in the boundary of U_σ. In the following, we no longer restrict to finite dimensions.

    Henceforth we assume that $I_\sigma^\perp \neq \{0\}$. As in Definition 3.3.6, given $\sigma \in B(G)$ with $\|\sigma\| = 1$, we fix a unitary $u \in I_\sigma^\perp$ such that $(I_\sigma^\perp, u)$ is a JW^*-algebra with identity u and there is an involution preserving isometry which carries $I_\sigma^\perp$ onto a JW^*-subalgebra $\mathcal{A}$ of $\mathcal{B}(H_u)$ for some complex Hilbert space H_u, and maps u to the identity 1_u in $\mathcal{B}(H_u)$. The Jordan product $\circ$ in $\mathcal{A}$ is given by

$$a \circ b = \frac{1}{2}(ab + ba)$$

and the triple product by

$$\{a, b, c\} = \frac{1}{2}(ab^*c + cb^*a)$$

where the product and involution in the right-hand side are those in $\mathcal{B}(H_u)$. We will identify $I_\sigma^\perp$ with $\mathcal{A}$ in the sequel and work with the latter.

An element $p \in \mathcal{A}$ is a *projection* if $p = p^* = p^2$. Let $\mathcal{P}$ be the set of all projections in $\mathcal{A}$, equipped with the relative topology. Then $\mathcal{P}$ is contained in the boundary of U_σ and is the union of its connected components. We study below the geometry of these components. We note that, in $I_\sigma^\perp$, $\mathcal{P}$ identifies with the set $\{p \in I_\sigma^\perp : p = P_\sigma(up^*u) = P_\sigma(pu^*p)\} \subset VN(G)$, here the adjoints u^* and p^* are taken in $VN(G)$.

Let $p \in \mathcal{P}$. The operator $p \,\square\, p : \mathcal{A} \to \mathcal{A}$ has eigenvalues $0, \frac{1}{2}, 1$ and we have the following *Peirce decomposition* of $\mathcal{A}$:

$$\mathcal{A} = \mathcal{A}_2(p) \oplus \mathcal{A}_1(p) \oplus \mathcal{A}_0(p)$$

where $\mathcal{A}_k(p)$ is the $\frac{k}{2}$-eigenspace for $k = 0, 1, 2$. The projection $P_k(p) : \mathcal{A} \to \mathcal{A}_k(p)$ is called the *Peirce k-projection* and is given by

$$P_2(p)(a) = pap$$

$$P_1(p)(a) = pa + ap - 2pap$$

$$P_0(p)(a) = (1_u - p)a(1_u - p).$$

Let $\mathcal{A}_1(p)_s = \{a \in \mathcal{A} : a^* = a = ap + pa\}$ be the self-adjoint part of $\mathcal{A}_1(p)$. For $a \in \mathcal{A}_1(p)_s$, we define the linear map $k_a = 2(a \,\square\, p - p \,\square\, a) : \mathcal{A} \to \mathcal{A}$, where $(a \square p)(\cdot) = \{a, p, \cdot\}$. Then k_a is a Jordan $*$-derivation :

$$k_a(x^*) = k_a(x)^*, \quad k_a(x \circ y) = x \circ k_a(y) + k_a(x) \circ y$$

for $x, y \in \mathcal{A}$. Therefore $\exp t k_a = \sum_{n=0}^{\infty} \frac{t^n k_a^n}{n!}$ is a Jordan $*$-automorphism of $\mathcal{A}$ for $t \in \mathbb{R}$.

Let $\mathcal{M}$ be a connected component of $\mathcal{P}$. Then $(\exp k_a)(\mathcal{M}) = \mathcal{M}$ and it has been shown in [16] that $\mathcal{M}$ is a real analytic Banach manifold and the local chart at a point $p \in \mathcal{M}$ is given by the map $a \in \mathcal{A}_1(p)_s \mapsto \exp k_a(p) \in \mathcal{M}$, where $\mathcal{A}_1(p)_s$ identifies with the tangent space $T_p\mathcal{M}$ at p. Further, $\mathcal{M}$ is a symmetric manifold with symmetry at $p \in \mathcal{M}$ given by $S_p = Id - 2P_1(p)$: $\mathcal{M} \rightarrow \mathcal{M}$.

Lemma 3.3.18. *The manifold* $\mathcal{M}$ *is path-connected.*

Proof. Given $p \in \mathcal{M}$, let

$$K_p = \{q \in \mathcal{M} : q \text{ can be joined to } p \text{ by a continuous path}\}.$$

Using the local chart at p, it can be shown easily that K_p is open. But $\mathcal{M} \backslash K_p = \underset{q \notin K_p}{\cup} K_q$ is open. So $\mathcal{M} = K_p$ by connectedness of $\mathcal{M}$. $\square$

Remark 3.3.19. It follows from the above lemma and [82; Proposition 5.2.10] that the projections in $\mathcal{M}$ are unitarily equivalent in $\mathcal{B}(H_u)$.

We next define an affine connection on $\mathcal{M}$ and describe the geodesics in $\mathcal{M}$. We fix some notation first. Let $T\mathcal{M}$ be the tangent bundle which is the disjoint union $\underset{p \in \mathcal{M}}{\cup} T_p\mathcal{M}$ of tangent spaces. A *vector field* on $\mathcal{M}$ is a section of $T\mathcal{M}$, that is, a map $X : \mathcal{M} \rightarrow T\mathcal{M}$ such that $X(p) \in T_p\mathcal{M}$. Since $T_p\mathcal{M}$ identifies with $\mathcal{A}_1(p)_s \subset \mathcal{A}$, we can think of X as an $\mathcal{A}$-valued map and we denote by $\mathcal{X}(\mathcal{M})$ the space of real analytic vector fields on $\mathcal{M}$. Given $X \in \mathcal{X}(\mathcal{M})$ and $f \in C^\infty(\mathcal{M})$, we define

$$(Xf)(p) = \frac{d}{dt} f(p + tX(p))\big|_{t=0}.$$

Note that fX denotes the vector field $(fX)(p) = f(p)X(p)$. An *affine connection* on $\mathcal{M}$ is a map

$$\nabla : \mathcal{X}(\mathcal{M}) \times \mathcal{X}(\mathcal{M}) \rightarrow \mathcal{X}(\mathcal{M})$$

with $\nabla(X, Y)$, usually written $\nabla_X Y$, satisfying the following conditions:

(i) $\nabla_X(Y_1 + Y_2) = \nabla_X Y_1 + \nabla_X Y_2$
(ii) $\nabla_{fX_1 + gX_2} Y = f\nabla_{X_1} Y + g\nabla_{X_2} Y$
(iii) $\nabla_X(fY) = f\nabla_X Y + (Xf)Y$.

Definition 3.3.20. Given $X, Y \in \mathcal{X}(\mathcal{M})$, we define

$$(\nabla_X Y)(p) = P_1(p)\big(Y'(p)(X(p))\big)$$

where $Y'(p) : \mathcal{A} \to \mathcal{A}$ is the Fréchet derivative of Y at p, and $P_1(p) : \mathcal{A} \to \mathcal{A}_1(p)$ is the Peirce 1-projection.

It has been shown in [16] that ∇ is a torsion free affine connection on $\mathcal{M}$.

Proposition 3.3.21. *Let $\mathcal{M}$ be a connected component of projections in $(I_\sigma^\perp, u)$. Given $p \in \mathcal{M}$ and a tangent vector $v \in T_p\mathcal{M}$, the ∇-geodesic at p with initial tangent vector v is given by*

$$\gamma_{p,v}(t) = (\exp tk_v)(p) \quad (t \in \mathbb{R})$$

where $k_v = 2(v \,\Box\, p - p \,\Box\, v)$.

Proof. We identify $I_\sigma^\perp$ with $\mathcal{A} \subset \mathcal{B}(H_u)$ as above and we do the computation in $\mathcal{A}$. Write $\gamma(t) = (\exp tk_v)(p)$. We need to show

$$(\nabla_{\dot\gamma} \dot\gamma)(\gamma(t)) = 0.$$

We have $\dot\gamma(t) = (\exp tk_v)\big(k_v(p)\big)$ and

$$\ddot\gamma(t) = (\exp tk_v)\big(k_v^2(p)\big) = (\exp tk_v)\big(k_v(v)\big)$$

where $v = pv + vp$ gives $k_v(p) = v - pvp$ and $k_v^2(p) = k_v(v) = 2(vpv - pv^2p)$. By Definition 3.3.20, we have

$$(\nabla_{\dot\gamma} \dot\gamma)(\gamma(t)) = P_1\big(\gamma(t)\big(\big(\dot\gamma(t)'(\gamma(t))\big)(\dot\gamma(t))\big)$$

$$= P_1\big(\gamma(t)\big)\big(\ddot\gamma(t)\big)$$

$$= P_1\big((\exp tk_v)(p)\big)(\exp tk_v)\big(k_v(v)\big).$$

Since $\exp tk_v$ is a triple automorphism of $\mathcal{A}$, the last term above is equal to $(\exp tk_v)P_1(p)\big(k_v(v)\big)$ and direct calculation gives $P_1(p)\big(k_v(v)\big) = 0$ which completes the proof. $\qquad\qquad\square$

In case $\mathcal{M}$ contains a rank-one projection in $\mathcal{B}(H_u)$, we have a more explicit description of the geodesics. Let $p \in \mathcal{M}$ be rank-one, that is, there exists a unit vector $\xi \in H_u$ such that

$$p(\cdot) = (\cdot|\xi)\,\xi$$

where $(\cdot|\cdot)$ is the inner product in H_u. By Remark 3.3.19, every projection in $\mathcal{M}$ is also rank-one in $\mathcal{B}(H_u)$. We can define a map $v \in T_p\mathcal{M} = A_1(p)$, $\mapsto$ $v(\xi) \in \{\xi\}^\perp$ where $v = vp + pv$ implies that $v(\xi)$ is in the orthogonal complement $\{\xi\}^\perp$ of $\{\xi\}$. Further, the above map is a linear homeomorphism, not necessarily surjective, as simple computation gives

$$\|v(\xi)\| \leq \|v\| \leq 2\|v(\xi)\|.$$

In fact, we have $\|v(\xi)\| = \|vp\| = \|pv\|$. It follows that $T_p\mathcal{M}$ is a *real* Hilbert space in the inner product

$$\langle\langle v_1, v_2\rangle\rangle_p = 2\,\mathrm{Re}\,(\,v_1(\xi)|v_2(\xi)\,).$$

We can therefore define a Riemannian metric g on $\mathcal{M}$ by

$$g(X,Y)_p = \langle\langle X(p), Y(p)\rangle\rangle_p$$

for $X, Y \in \mathcal{X}(\mathcal{M})$ and $p \in \mathcal{M}$. As in [16], one can show that ∇ is compatible with g, and is therefore the Levi-Civita connection on $\mathcal{M}$. Also, using similar arguments involving Jordan arithmetic as in [16], the ∇-geodesic at $p \in \mathcal{M}$ with initial tangent vector $v \in T_p\mathcal{M}$ can be expressed as

$$\gamma_{p,v}(t) = \cos(2\|pv\|t)p + \frac{\sin(2\|pv\|t)}{2\|pv\|}\,v + \frac{1 - \cos(2\|pv\|t)}{2\|pv\|^2}\,v^2$$

which is contained in the closed real Jordan subalgebra of $(I_\sigma^\perp, u)$ generated by p and v. One can deduce that the Riemann distance between $p, q \in \mathcal{M}$ is $\sqrt{2}\,\sin^{-1}\|p - q\|$.

3.4. Classification of harmonic functionals

The Murray-von Neumann classification of rings of operators [62] is fundamental in the theory of operator algebras and has been extended to Jordan algebras [79] and Jordan triples [43,44].

In this section, we study the Murray-von Neumann classification of the harmonic functionals $I_\sigma^\perp$ in terms of the linear geometry of its predual $A(G)/I_\sigma$. We show that the classification of $I_\sigma^\perp$ is completely determined by the facial structures of its normal state space in $A(G)/I_\sigma$. By fixing a unitary element $u \in I_\sigma^\perp$ as before, we can identify $(I_\sigma^\perp, u)$ with a JW^*-subalgebra $\mathcal{A}$ of some $\mathcal{B}(H)$ containing the identity operator $\mathbf{1} : H \to H$. We will show that the classification of $I_\sigma^\perp$ does not depend on the choice of u. In fact, we prove our classification results for the general case of arbitrary JW^*-algebras and obtain the results for $I_\sigma^\perp$ as special case.

We begin with the classification of JW^*-algebras. Throughout $\mathcal{A}$ denotes a JW^*-subalgebra of the algebra $\mathcal{B}(H)$ of bounded linear operators on some Hilbert space H in which the Jordan product $\circ$ is defined by

$$a \circ b = \frac{1}{2}(ab + ba)$$

and the Jordan triple product is given by

$$\{a, b, c\} = \frac{1}{2}(ab^*c + cb^*a)$$

and also, the identity of $\mathcal{A}$ is the identity operator $\mathbf{1} : H \to H$. Let

$$A = \{a \in \mathcal{A} : a^* = a\}$$

be the *self-adjoint part* of $\mathcal{A}$. Then $\mathcal{A} = A + iA$ and $(A, \circ)$ is a real Jordan Banach algebra, called a JW-algebra. Two elements $a, b \in A$ are said to *operator commute* if $T_a T_b = T_b T_a$ where $T_a : A \to A$ is defined by $T_a(x) = a \circ x$ and T_b is defined likewise. Clearly $T_a T_b = T_b T_a$ is equivalent to $a \circ (b \circ c) = (a \circ c) \circ b$ for all $c \in A$. The *centre* $\mathcal{Z}$ of $\mathcal{A}$ is defined by

$$\mathcal{Z} = \{a \in \mathcal{A} : a \circ (x \circ y) = (a \circ y) \circ x \quad \text{for all } x, y \in \mathcal{A}\}.$$

Its self-adjoint part Z is the centre of A. $\mathcal{A}$ is called a *factor* if $\mathcal{Z} = \mathbb{C}\mathbf{1}$. We note that $\mathcal{A}$ is uniformly generated by projections which determine its classification.

A projection $p \in \mathcal{A}$ is called *abelian* if $(p\mathcal{A}p, \circ)$ is an associative Jordan algebra in which case it is an abelian von Neumann algebra. Two projections $p, q \in \mathcal{A}$ are *equivalent*, denoted by $p \sim q$, if there exist $s_1, \ldots, s_n \in A$ such that $s_j^2 = 1$ for $j = 1, \ldots, n$ and $s_n \ldots s_1 p s_1 \ldots s_n = q$. By [79; p. 23], we have $p \sim q$ if, and only if, $1 - p \sim 1 - q$. We equip A with the usual ordering $\leq$ in $\mathcal{B}(H)$. A JW^*-algebra $\mathcal{A}$ is called *modular* if every projection $p \in \mathcal{A}$ has the following property:

$$\text{given a projection} \quad q \leq p \quad \text{with} \quad q \sim p, \text{ then } \quad q = p$$

(cf. [79; Proposition 14]). A projection $p \in \mathcal{A}$ is called *modular* if $(p\mathcal{A}p, \circ)$ is modular. We note that abelian projections are modular. The projections in the centre are called the *central projections*. $\mathcal{A}$ is called *properly non-modular* if it has no nonzero modular central projection.

A JW^*-algebra $\mathcal{A}$ is of *type I* if every nonzero central projection in $\mathcal{A}$ dominates a nonzero abelian projection; $\mathcal{A}$ is of *type* II if it has no nonzero abelian projection and every nonzero central projection in $\mathcal{A}$ dominates a nonzero modular projection; $\mathcal{A}$ is of *type* III if it has no nonzero modular projection. A *type* II_1 JW^*-algebra is one that is type II and modular. A type II and properly non-modular JW^*-algebra is said to be of *type* II_∞. Our definition of types for $\mathcal{A} = A + iA$ coincides with that for the JW-algebra A given in [79].

It is not obvious that the types of JW^*-algebras are invariant under linear isometries although this will be an immediate consequence of our characterization of types in terms of the geometry of the normal state space. We note that it has been shown in [44; p. 57] that linear isometries preserve types II and III, using the nontrivial fact that JW^*-algebras of these types generate von Neumann algebras of the same types. Our unified geometric approach does not require this fact.

As in [79; Theorem 13], one can show that every JW^*-algebra decomposes uniquely into direct summands of types I, II_1, II_∞ and III. Complexifying the result in [79; Theorem 26], one has the following important characterization of modular JW^*-algebras.

Lemma 3.4.1. *A JW^*-algebra $\mathcal{A}$ is modular if, and only if, it has a unique faithful normal centre-valued trace, that is, there is a unique w^*-continuous linear map $\sharp : \mathcal{A} \to \mathcal{Z}$ satisfying the following conditions:*

(i) $\sharp(a) \geq 0$ *for* $a \geq 0$;
(ii) $\sharp(a) = 0$ *and* $a \geq 0$ *imply* $a = 0$;
(iii) $\sharp(a \circ z) = \sharp(a) \circ z$ *for* $z \in \mathcal{Z}$;

(iv) $\sharp(sa^*s) = \sharp(a)^*$ *for* $s \in A$ *with* $s^2 = 1$

(v) $\sharp(1) = 1$.

We note that $a \circ z = az$ for $z \in \mathcal{Z}$. We also note that a von Neumann algebra $\mathcal{A}$ is finite if, and only if, $(\mathcal{A}, \circ)$ is a modular JW^*-algebra with Jordan product $a \circ b = \frac{1}{2}(ab + ba)$. The type of a von Neumann algebra $\mathcal{A}$ is the same as that of the JW^*-algebra $(\mathcal{A}, \circ)$. Obviously, even if $I_\sigma^\perp$ is a von Neumann subalgebra of $VN(G)$, it need not inherit the type of $VN(G)$, for instance, they could have different dimensions. However, if $I_\sigma^\perp$ is a subalgebra of $VN(G)$, then finiteness of $VN(G)$ passes onto $I_\sigma^\perp$ and the image of the centre of $VN(G)$ under $P_\sigma : VN(G) \to I_\sigma^\perp$ is contained in the centre of $I_\sigma^\perp$ by Remark 3.3.8. Further, if $\sigma^{-1}\{1\}$ is an open subgroup of G, then Corollary 3.3.4 implies that P_σ is w^*-continuous in which case $I_\sigma^\perp$ is type I if $VN(G)$ is, and if $VN(G)$ contains no type III summand, then $I_\sigma^\perp$ does not contain such either (cf. [78]).

Example 3.4.2. If G is a countable discrete group in which every conjugacy class, except the identity class, is infinite, then $VN(G)$ is a type II$_1$ factor. The group G_p of all finite permutations of an infinite countable set is such a group and as in Section 3.3, we can find $\sigma \in B(G_p)$ such that $I_\sigma^\perp$ is a 2-dimensional abelian subalgebra of $VN(G_p)$.

We note that $VN(G)$ is finite if, and only if, G is a [SIN]-group. We next characterize the modularity of $I_\sigma^\perp$ by its normal state space.

We recall that a continuous linear functional on a JBW^*-triple Z is *normal* if it is additive on orthogonal tripotents, that is, $f(\sum_\alpha e_\alpha) = \sum_\alpha f(e_\alpha)$ for any orthogonal family $\{e_\alpha\}$ of tripotents in Z where the sum is taken with respect to the w^*-topology $\sigma(Z, Z_*)$ and two tripotents e_α, e_β are *orthogonal* if $e_\alpha \Box e_\beta = 0$. The w^*-continuous linear functionals on Z are exactly the normal functionals [42; Proposition 3.19]. In the case of JW^*-algebras, the normal functionals are exactly the ones which are additive on orthogonal projections [42; Proposition 3.18]. A continuous linear functional f of a JW^*-algebra $\mathcal{A}$ is called a *state* if $\|f\| = 1 = f(1)$. Let

$$N = \{f \in \mathcal{A}_* : \|f\| = 1 = f(1)\}$$

be the normal state space of $\mathcal{A}$. It is a norm-closed convex subset of $\mathcal{A}_*$. A face F of N is called a *split face* if there is a face F' of N, disjoint from F, such that each $f \in N\backslash(F \cup F')$ has a unique representation

$$f = \alpha g + (1 - \alpha)h$$

where $0 < \alpha < 1$, $g \in F$ and $h \in F'$. Split faces are norm-closed. Given a projection $p \in \mathcal{A}$, the set $F = \{f \in N : \langle f, p \rangle = 1\}$ is a norm-closed face of N, and all norm-closed faces of N are of this form. Moreover, F identifies with the normal state space of $p\mathcal{A}p$ and it is a split face if, and only if, p is a central projection (cf. [3; Theorem 11.5] and [2; Proposition 4.11]). We define an *affine symmetry* of F to be an affine bijection $\zeta : F \to F$ such that $\zeta(K) = K$ for every split face K of F. We will denote by $\mathrm{Sym}\,(F)$ the set of all affine symmetries of F.

A linear functional f on $\mathcal{A} = A + iA$ is *self-adjoint* if $f = f^*$ where $f^* : \mathcal{A} \to \mathbb{C}$ is defined by $f^*(a) = \overline{f(a)}$. The self-adjoint part of $\mathcal{A}_*$, denoted by $A_* = \{f \in \mathcal{A}_* : f^* = f\}$, is a real Banach space partially ordered by the cone $C = \bigcup_{\alpha \geq 0} \alpha N$ and is the predual of A. We have $A_* = C - C$ and the norm of A_* is additive on C. Given an affine bijection $\zeta : N \to N$, we can extend it naturally to a linear order-isomorphism $\widetilde{\zeta} : A_* \to A_*$. Further, $\widetilde{\zeta}$ is an isometry since (A_*, C) is a base norm space with base N which implies that for each $f \in A_*$, we have

$$\|f\| = \inf\{\alpha + \beta : f = \alpha g - \beta h, \alpha, \beta \geq 0, g, h \in N\}$$

(cf. [1; p. 36]). Let $\zeta^* : A \to A$ be the dual map of $\widetilde{\zeta}$. Then ζ^* is a surjective linear isometry and $\zeta^*(1) = 1$. By [84; Theorem 4], ζ^* is a Jordan automorphism. It follows that the complexified map $\Phi_\zeta = \zeta^* + i\zeta^* : \mathcal{A} \to \mathcal{A}$ is a Jordan $*$-automorphism.

A linear functional f on $\mathcal{A}$ is called a *trace* if it satisfies $f(a) = f(sas)$ for all $a \in \mathcal{A}$ and $s \in \mathcal{A}$ with $s^2 = 1$. If $\mathcal{A}$ is modular with centre-valued trace $\sharp : \mathcal{A} \to \mathcal{Z}$, then for any linear functional ρ of $\mathcal{Z}$, the composite $\rho \circ \sharp$ is clearly a trace of $\mathcal{A}$. In fact, every trace f of $\mathcal{A}$ is of this form. To see this, let $\rho = f|_{\mathcal{Z}}$ be the restriction of f to $\mathcal{Z}$. Given $a \in A$, let K_a be the norm-closed convex hull of $\{s_n \ldots s_1 a s_1 \ldots s_n : s_j \in A, s_j^2 = 1\}$ in A. Then by [79; Lemma 31], $K_a \cap Z = \{\sharp(a)\}$. Since $f(K_a) = \{f(a)\}$, we have $(\rho \circ \sharp)(a) = f(\sharp(a)) = f(a)$.

Lemma 3.4.3. *Let $\mathcal{A}$ be a modular JW^*-algebra with normal state space N. Let $f \in N$ be a trace. Then f is a fixed-point of the symmetries of N.*

Proof. Let $\zeta : N \to N$ be a symmetry and let $\Phi_\zeta = \zeta^* + i\zeta^* : \mathcal{A} \to \mathcal{A}$ be the Jordan $*$-automorphism defined above. Let $\sharp : \mathcal{A} \to \mathcal{Z}$ be the unique centre-valued trace. We have $f = \rho \circ \sharp$ where $\rho = f|_{\mathcal{Z}}$ is a normal state of $\mathcal{Z}$. We first show that Φ_ζ fixes the centre of $\mathcal{Z}$. Since $\mathcal{Z}$ is uniformly generated by its projections, it suffices to show that $\Phi_\zeta(p) = p$ for all projections p in

$\mathcal{Z}$. Let $F = \{f \in N : \langle f,p\rangle = 1\}$. Then F is a split face of N as remarked before. The complementary face of F is $F' = \{f \in N : \langle f,p\rangle = 0\}$. For every $f \in N\backslash(F \cup F')$, we have

$$f = \alpha g + (1-\alpha)h$$

where $0 < \alpha < 1$, $g \in F$ and $h \in F'$. Since ζ leaves F and F' invariant, we have $\zeta(g) \in F$ and $\zeta(h) \in F'$ which gives

$$\langle f,p\rangle = \langle \alpha g + (1-\alpha)h, p\rangle = \alpha = \langle \alpha\zeta(g) + (1-\alpha)\zeta(h), p\rangle$$

$$= \langle \zeta(f), p\rangle = \langle f, \zeta^*(p)\rangle = \langle f, \Phi_\zeta(p)\rangle.$$

This proves $\Phi_\zeta(p) = p$. Next we show that $\sharp \circ \Phi_\zeta : \mathcal{A} \to \mathcal{Z}$ is a faithful normal centre-valued trace. Since Φ_ζ preserves the Jordan product and leaves $\mathcal{Z}$ fixed, we have, for $a \in \mathcal{A}$ and $z \in \mathcal{Z}$,

$$(\sharp \circ \Phi_\zeta)(a \circ z) = \sharp(\Phi_\zeta(a) \circ \Phi_\zeta(z)) = \sharp(\Phi_\zeta(a) \circ z) = \sharp(\Phi_\zeta(a)) \circ z.$$

For $s \in \mathcal{A}$ and $s^2 = 1$, we have $\Phi_\zeta(s)^2 = \Phi_\zeta(s^2) = 1$ and

$$(\sharp \circ \Phi_\zeta)(sas) = \sharp(\Phi_\zeta(s)\Phi_\zeta(a)\Phi_\zeta(s)) = \sharp(\Phi_\zeta(a)).$$

It follows that $\sharp \circ \Phi_\zeta$ is a faithful normal centre-valued trace and by uniqueness of $\sharp$, we have $\sharp \circ \Phi_\zeta = \sharp$ which implies that, for $a \in \mathcal{A}$,

$$\langle \zeta(f), a\rangle = \langle f, \Phi_\zeta(a)\rangle = \langle \rho \circ \sharp, \Phi_\zeta(a)\rangle = \rho(\sharp \circ \Phi_\zeta(a))$$

$$= \langle \rho \circ \sharp, a\rangle = \langle f, a\rangle$$

that is, $\zeta(f) = f$. $\qquad\qquad\qquad\qquad\qquad\qquad\qquad\qquad\qquad\square$

Modularity of a JW^*-algebra $\mathcal{A}$ can be characterized by the geometric property that there are sufficiently many fixed points of the affine symmetries of its normal state space N. To make this precise, a subset S of a norm-closed face $F = \{f \in N : \langle f,p\rangle = 1\} \subset N$ is said to be *separating in F* if for any nonzero positive $a \in p\mathcal{A}p$, there exists $f \in S$ such that $\langle f,a\rangle > 0$. 'Sufficiently many fixed-points' means that the fixed-points are separating in N.

Lemma 3.4.4. *Let A be a JW^*-algebra with normal state space N. Then A is modular if, and only if, the set of fixed-points of the affine symmetries of N is separating.*

Proof. Let A be modular with the faithful normal centre-valued trace $\sharp : A \to Z$. The set

$$\{\rho \circ \sharp : \rho \text{ is a normal state of } Z\} \subset N$$

is separating in N and by Lemma 3.4.3, each $\rho \circ \sharp$ is a fixed-point of $\zeta \in \text{Sym}(N)$.

To prove the converse, we first note that every fixed-point f of $\text{Sym}(N)$ is a trace. Indeed, given $s \in A$ with $s^2 = 1$, the map $\zeta : N \to N$ defined by $\zeta(h)(\cdot) = h(s \cdot s)$ is an affine symmetry since for any central projection $p \in A$, $\langle h, p \rangle = 1$ implies $\langle \zeta(h), p \rangle = \langle h, sps \rangle = 1$, in other words, $\zeta\{h \in N : \langle h, p \rangle = 1\} = \{h \in N : \langle h, p \rangle = 1\}$. Hence $\zeta(f) = f$ gives $f(a) = f(sas)$ for all $a \in A$. For each $a \in A$, let K_a be the norm-closed convex hull of $\{s_n \ldots s_1 a s_1 \ldots s_n : s_j \in A, s_j^2 = 1\}$ in A. By [79; Theorem 25], $K_a \cap Z$ is non-empty. To complete the proof, we need only show that $K_a \cap Z$ is a singleton which will then define the required centre-valued trace on A. Suppose there exist $x, y \in K_a \cap Z$ and $x \neq y$. Representing Z as continuous functions on a compact Hausdorff Stonean space, we can easily find a projection $q \in Z$ and $\alpha > 0$ such that either $xq > yq + \alpha q$ or $yq > xq + \alpha q$. Assuming the former say, then there is, by assumption, a fixed point f of $\text{Sym}(N)$ such that $f(xq) > f(yq) + \alpha f(q)$. But $f(xq) = f(qxq) = f(x) = f(a) = f(y) = f(yq)$ which gives a contradiction. So $K_a \cap Z$ is a singleton and the proof is complete. $\square$

It is well-known that a JW^*-algebra A is associative if, and only if, its normal state space N is *simplicial*, that is, N is a simplex: given any $\alpha \geq 0$ and $f \in A_*$, $(\alpha N + f) \cap N$ is either empty or of the form $\beta N + g$ for some $\beta \geq 0$ and $g \in A_*$. Given two projections $p, q \in A$, it is clear that $p \leq q$ if, and only if,

$$\{f \in N : \langle f, p \rangle = 1\} \subset \{f \in N : \langle f, q \rangle = 1\}.$$

Now putting the above results and remarks together, we obtain the following characterization of types of A in terms of the facial structures of the normal state space N. The following extends the main result in [12].

Theorem 3.4.5. *Let A be a JW^*-algebra with normal state space N. Then we have*

(i) A *is of type I if, and only if, every split face of N contains a simplicial face;*

(ii) A *is of type II_1 if, and only if, N has no simplicial face and the fixed points of* $\mathrm{Sym}\,(N)$ *are separating;*

(iii) A *is of type II_∞ if, and only if, N has no simplicial face and $\mathrm{Sym}\,(F)$ has no fixed-point for every split face $F \subset N$, but F contains a norm-closed face F_1 in which the fixed-points of* $\mathrm{Sym}\,(F_1)$ *are separating;*

(iv) A *is of type III if, and only if, $\mathrm{Sym}\,(F)$ has no fixed-point for every norm-closed face F of N.*

In particular, the result holds for $I_\sigma^\perp$ and its normal state space for $\sigma \in B(G)$ and $\|\sigma\| = 1$.

Proof. We have (ii) by Lemma 3.4.4. For (iii), we only need to show that A has no modular central projection if, and only if, $\mathrm{Sym}\,(F)$ has no fixed-point for every split face $F \subset N$. Indeed, if A has a nonzero modular central projection z, then zAz is modular with normal state space $N_z = \{f \in N : \langle f, z \rangle = 1\}$ which is a split face of N and Lemma 3.4.3 implies that $\mathrm{Sym}\,(N_z)$ has a fixed-point. Conversely, given a split face $F = \{f \in N : \langle f, p \rangle = 1\}$ for some central projection $p \in A$, if $\mathrm{Sym}\,(F)$ has a fixed-point $f_0 \in F$, then f_0 is a trace of $pAp(= Ap)$ as in the proof of Lemma 3.4.4. There is a smallest central projection $q \in A$, called the *central support* of f_0 (cf. [2; §5]) such that $\langle f_0, q \rangle = 1$. Since $F_q = \{f \in N : \langle f, q \rangle = 1\}$ is a split face in F, an affine symmetry of F_q extends to one of F. Since f_0 is faithful on qAq, qAq is modular by Lemma 3.4.4, that is, q is a modular central projection in A. Alternatively, modularity of qAq follows from the fact that a JW^*-algebra is modular if it admits a faithful normal trace.

Similar arguments apply to (iv) by observing the correspondence between the norm-closed faces of N and the projections in A, and using the support projection (cf. [2; §5]) instead of the central support of a fixed-point f_0 of $\mathrm{Sym}\,(F)$ as in the proof of (iii). $\qquad\square$

Corollary 3.4.6. *Let $\Phi : A \to B$ be a surjective linear isometry between two JW^*-algebras A and B. Then A and B have the same type.*

Proof. By uniqueness of predual, Φ is w^*-w^*-continuous (cf. [42; Corollary 3.22]) and induces a norm continuous affine bijection $\Phi_* : N_B \to N_A$ between the normal state spaces of B and A, and by Theorem 3.4.5, Φ preserves the types. $\qquad\square$

Corollary 3.4.7. *Let* $\sigma \in B(G)$ *and* $\|\sigma\| = 1$. *Let* $u, v \in I_\sigma^\perp$ *be unitaries in* $VN(G)$. *Then* $(I_\sigma^\perp, u)$ *and* $(I_\sigma^\perp, v)$ *have the same type.*

Finally we note that the type structures of $I_\sigma^\perp$ are also related to some interesting Banach space properties of $A(G)/I_\sigma$. By [19; Theorem 2], we have that $I_\sigma^\perp$ is an ℓ_∞-direct sum of type I factors if, and only if, $A(G)/I_\sigma$ has the Radon-Nikodym property. Also, by [20; Theorem 20], $I_\sigma^\perp$ is type I modular if, and only if, $A(G)/I_\sigma$ has the Dunford-Pettis property.

REFERENCES

[1] L. Asimow and A.J. Ellis, Convexity theory and its applications in functional analysis, Academic Press, London, 1981.

[2] E.M. Alfsen and F.W. Schultz, *A Gelfand-Neumark theorem for Jordan algebras*, Adv. in Math. **28** (1978), 11-56.

[3] E.M. Alfsen and F.W. Schultz, *Non-commutative spectral theory for affine function spaces*, Memoirs Amer. Math. Soc. 172 (1976).

[4] A. Avez, *Théorème de Choquet-Deny pour les groupes à croissance non-exponentielle*, C.R. Acad. Sc. **279** (1974), 25-28.

[5] R. Azencott, *Espaces de Poisson des groupes localement compacts*, Lecture Notes in Math. 148, Springer-Verlag, Berlin, 1970.

[6] J. Baker, A.T. Lau and J. Pym, *Module homomorphisms and topological centres associated with weakly sequentially complete Banach algebras*, J. of Funct. Anal. **158** (1998), 186-208.

[7] T.J. Barton, T. Dang and G. Horn, *Normal representations of Banach Jordan triple systems*, Proc. Amer. Math. Soc. **102** (1988), 551-556.

[8] J.F. Berglund, H.D. Junghenn and P. Milnes, *Analysis on semigroups*, Canadian Math. Soc. Series of Monographs and Adv. Texts, J. Wiley & Sons, New York, 1989.

[9] R.B. Burkel, *Weakly almost periodic functions on semigroups*, Gordon and Breach, New York, 1970.

[10] E. Cartan, *Sur les domaines bornés homogènes de l'espace de n variables complexes*, Abh. Math. Semin. Univ. Hamburg **11** (1935), 116-162.

[11] G. Choquet and J. Deny, *Sur l'équation de convolution* $\mu = \mu * \sigma$, C.R.

Acad. Sc. Paris **250** (1960), 779-801.

[12] C-H. Chu, *Remarks on the classification of von Neumann algebras*, Operator Algebras and Operator Theory, Pitman Res. Notes 271, 1992, pp. 62-68.

[13] C-H. Chu, *Jordan structures in Banach manifolds*, Studies in Adv. Math. 20, Amer. Math. Soc. (2001), 201 -210.

[14] C-H. Chu and T. Hilberdink, *The convolution equation of Choquet and Deny on nilpotent groups*, Integral Equations & Oper. Th. **26** (1996), 1-13.

[15] C-H. Chu, *Matrix-valued harmonic functions on groups*, J. Reine Angew. Math. (to appear).

[16] C-H. Chu and J.M. Isidro, *Manifolds of tripotents in JB^*-triples*, Math. Z. **233** (2000), 741-754.

[17] C-H. Chu and C-W. Leung, *Harmonic functions on homogeneous spaces*, Monatshefte Math. **128** (1999), 227-235.

[18] C-H. Chu and C-W. Leung, *The convolution equation of Choquet and Deny on [IN]-groups*, Integral Equations & Oper. Th. **40** (2001), 391-402.

[19] C-H. Chu and B. Iochum, *On the Radon-Nikodym property in Jordan triples*, Proc. Amer. Math. Soc. **99** (1987), 462-464.

[20] C-H. Chu and P. Mellon, *The Dunford-Pettis property in JB^*-triples*, J. London Math. Soc. **55** (1997), 515-526.

[21] J. Deny, *Sur l'équation de convolution $\mu * \sigma = \mu$*, Semin. Théo. Potentiel de M. Brelot, Paris, 1960.

[22] E.B. Dynkin and M.B. Malyutov, *Random walks on groups with a finite number of generators*, Soviet Math. Doklady **2** (1961), 399-402.

[23] R. Ellis, *Locally compact transformation groups*, Duke Math. J. **24** (1957), 119-125.

[24] P. Eymard, *L'algèbre de Fourier d'un groupe localement compact*, Bull. Soc. Math. France **92** (1964), 181-236.

[25] Y. Friedman and B. Russo, *Solution of the contractive projection problem*, J. Funct. Anal. **60** (1985), 67-89.

[26] Y. Friedman and B. Russo, *Contractive projections on $C_0(K)$*, Trans. Amer. Math. Soc. **273** (1982), 57-73.

[27] H.H. Furstenberg, *A Poisson formula for semi-simple Lie groups*, Ann. of Math. **77** (1963), 335-368.

[28] S. Glasner, *Proximal flows*, Lecture Notes in Math. 517, Springer-Verlag, Berlin, 1976.

[29] C.C. Graham, A.T. Lau and M. Leinert, *Separable translation invariant subspaces of $M(G)$ and other dual spaces on locally compact groups*, Colloq. Math. **IV** (1988), 131-145.

[30] E.E. Granirer, *When quotients of the Fourier algebra $A(G)$ are ideals in their bidual and when $A(G)$ has WCHP*, Math. Japonica **46** (1997), 69-72.

[31] E.E. Granirer, *On some properties of the Banach algebras $A_p(G)$ for locally compact groups*, Proc. Amer. Math. Soc. **95** (1985), 375-381.

[32] E.E. Granirer, *Density theorems for some linear subspaces and some C^*-subalgebras of $VN(G)$*, Symposia Math. **20** (1977), 61-70.

[33] E.E. Granirer, *Weakly almost periodic and uniformly continuous functionals on the Fourier algebra of a locally compact group*, Trans. Amer. Math. Soc. **189** (1974), 372-382.

[34] E. Granirer and M. Leinert, *On some topologies which coincide on the*

unit sphere of the Fourier-Stieltjes algebra $B(G)$ *and the measure algebra* $M(G)$, Rocky Mountain J. Math. **11** (1981), 459-472.

[35] S. Grosser, R. Mosak and M. Moskowitz, *Duality and harmonic analysis on central topological groups I and II*, (correction: ibid, p. 375), Indag. Math. **35** (1973), 65-91.

[36] Y. Guivarc'h, *Croissance polynomiale de périodes des fonctions harmoniques*, Bull. Soc. Math. France **101** (1973), 333-379.

[37] C.S. Herz, *Harmonic synthesis for subgroups*, Ann. Inst. Fourier **23** (1973), 91-123.

[38] E. Hewitt and K.A. Ross, *Abstract Harmonic Analysis*, vol. I, Springer-Verlag, Berlin, 1963.

[39] K.H. Hofmann, J.D. Lawson and J.S. Pym, *The analytical and topological theory of semigroups*, Walter de Gruyter, Berlin, 1990.

[40] K.H. Hofmann and P.S. Mostert, *Elements of compact semigroups*, C.E. Merrill Books, Ohio, 1966.

[41] B. Host, J.-F. Méla and F. Parreau, *Analyse harmonique des mesures*, Astérisque 135-136, Soc. Math. France (1986).

[42] G. Horn, *Characterization of the predual and ideal structure of a JBW*-triple*, Math. Scand. **61** (1987), 117-133.

[43] G. Horn, *Classification of JBW*-triples of type I*, Math. Z. **196** (1987), 271-291.

[44] G. Horn and E. Neher, *Classification of continuous JBW*-triples*, Trans. Amer. Math. Soc. **306** (1988), 553-578.

[45] B.E. Johnson, *Harmonic functions on nilpotent groups*, Integral Equations & Oper. Th. **40** (2001), 454-464.

[46]   V.A. Kaimanovich and A.M. Vershik, *Random walks on discrete groups: boundary and entropy*, Ann. of Prob. **11** (1983), 457-490.

[47]   E. Kaniuth and A.T Lau, *Spectral synthesis for A(G) and subspaces of VN(G)**, Proc. Amer. Math. Soc. **129** (2001), 3253-3263.

[48]   W. Kaup, *A Riemann mapping theorem for bounded symmetric domains in complex Banach spaces*, Math. Z. **144** (1975), 75-96.

[49]   J.L. Kelley, *General topology*, van Nostrand, New Jersey, 1986.

[50]   A.W. Knapp, *Distal functions on groups*, Trans. Amer. Math. Soc. **128** (1967), 1-40.

[51]   Y. Kwada and K. Ito, *On the probability distribution on a compact group I*, Proc. Phys. Math. Soc. Japan **22** (1940), 977-998.

[52]   A.T. Lau, *Analysis on a class of Banach algebras with applications to harmonic analysis on locally compact groups and semigroups*, Fund. Math. **118** (1983), 161-175.

[53]   A.T. Lau, *Uniformly continuous functionals on the Fourier algebra of any locally compact groups*, Trans. Amer. Math. Soc. **251** (1979), 39-59.

[54]   A.T. Lau, *Operators which commute with convolutions on subspaces of $L_\infty(G)$*, Colloq. Math. **39** (1978), 351-359.

[55]   A.T. Lau, *Semigroup of operators on dual Banach spaces*, Proc. Amer. Math. Soc. **54** (1976), 393-396.

[56]   A.T. Lau and V. Losert, *The C^*-algebra generated by operators with compact support on a locally compact group*, J. Funct. Anal. **112** (1993), 1-30.

[57]   A.T. Lau and J. Pym, *The topological centre of a compactification of a locally compact group*, Math. Z. **219** (1995), 567-579.

[58] H. Leptin, *Sur l'algèbre de Fourier d'un groupe localement compact*, C.R. Acad. Sci. Paris Sér A-B **266** (1968), A1180-A1182.

[59] J. Lindenstrauss and D.E. Wulbert, *On the classification of Banach spaces whose duals are L_1-spaces*, J. Func. Anal. **4** (1969), 332-349.

[60] K. McKennon, *Multipliers, positive functionals, positive definite functions and Fourier-Stieltjes transforms*, Memoirs Amer. Math. Soc. 111 (1971).

[61] P. Milnes, *Uniformity and uniformly continuous functions on locally compact groups*, Proc. Amer. Math. Soc. **109** (1990), 567-570.

[62] F. Murray and J. von Neumann, *On rings of operators*, Ann. of Math. **37** (1936), 116-229.

[63] A.L.T. Paterson, *Amenability*, Math. Surveys and Monographs 29, Amer. Math. Soc., 1988.

[64] A.L.T. Paterson, *A non-probabilistic approach to Poisson spaces*, Proc. Roy. Soc. Edinburgh **93A** (1983), 181-188.

[65] V.I. Paulsen, *Completely bounded maps and dilations*, Pitman Research Notes **146**, Longman, Essex, 1986.

[66] A. Raugi, *Fonctions harmoniques positives sur certains groupes de Lie resolubles connexes*, Bull. Soc. Math. France **124** (1996), 649-684.

[67] P.F. Renaud, *Invariant means on a class of von Neumann algebras*, Trans. Amer. Math. Soc. **170** (1972), 285-291.

[68] D. Revuz, *Markov chains*, North-Holland, Amsterdam, 1975.

[69] A.P. Robertson and W.J. Robertson, *Topological vector spaces*, Camb. Univ. Press, Cambridge, 1980.

[70] J. Rosenblatt, *Ergodic and mixing random walks on locally compact*

groups, Math. Ann. **257** (1981), 31-42.

[71]   W. Ruppert, *Rechtstopologische Halbgruppen*, J. Angew. Math. **261** (1973), 123-133.

[72]   B. Russo, *Structures of JB* triples*, Proceedings of the 1992 Oberwolfach Conference on Jordan Algebras, Walter de Gruyter, Berlin, 1994, pp. 208-280.

[73]   L. Schwartz, *Théorie génerale des fonctions moyennes-périodiques*, Ann. of Math. **48** (1947), 857-929.

[74]   M. Takesaki, *Theory of operator algebras 1*, Springer Verlag, New York, 1979.

[75]   M. Takesaki and N. Tatsuma, *Duality and subgroups II*, J. Funct. Anal. **11** (1972), 184-190.

[76]   M.E. Taylor, *Noncommutative harmonic analysis*, Amer. Math. Soc., Providence (1986).

[77]   J. Tomiyama, *On the projection of norm one in W^*-algebras*, Proc. Japan Acad. **33** (1957), 608-612.

[78]   J. Tomiyama, *On the projection of norm one in W^*-algebras, III*, Tôhoku Math. J. **11** (1959), 125-129.

[79]   D.M. Topping, *Jordan algebras of self-adjoint operators*, Memoirs Amer. Math. Soc. 53 (1953).

[80]   H. Upmeier, *Symmetric Banach Manifolds and Jordan C^*-Algebras*, North-Holland, Amsterdam, 1985.

[81]   M. E. Walter, *A duality between locally compact groups and certain Banach algebras*, J. Funct. Anal. **17** (1974), 131-160.

[82]   N.E. Wegge-Olsen, *K-theory and C^*-algebras*, Oxford Univ. Press, Ox-

ford, 1993.

[83] G.A. Willis, *Probability measures on groups and some related ideals in group algebras*, J. Funct. Anal. **92** (1990), 202-263.

[84] J.D. Maitland Wright and M. A. Youngson, *On isometries of Jordan algebras*, J. London Math. Scoc. **17** (1978), 339-344.

LIST OF SYMBOLS

INDEX

Lecture Notes in Mathematics

For information about Vols. 1–1593
please contact your bookseller or Springer-Verlag

gation. Montecatini Terme, 1994. Editor: T. Ruggeri. VII, 142 pages. 1996.

Vol. 1641: P. Abramenko, Twin Buildings and Applications to S-Arithmetic Groups. IX, 123 pages. 1996.

Vol. 1642: M. Puschnigg, Asymptotic Cyclic Cohomology. XXII, 138 pages. 1996.

Vol. 1643: J. Richter-Gebert, Realization Spaces of Polytopes. XI, 187 pages. 1996.

Vol. 1644: A. Adler, S. Ramanan, Moduli of Abelian Varieties. VI, 196 pages. 1996.

Vol. 1645: H. W. Broer, G. B. Huitema, M. B. Sevryuk, Quasi-Periodic Motions in Families of Dynamical Systems. XI, 195 pages. 1996.

Vol. 1646: J.-P. Demailly, T. Peternell, G. Tian, A. N. Tyurin, Transcendental Methods in Algebraic Geometry. Cetraro, 1994. Editors: F. Catanese, C. Ciliberto. VII, 257 pages. 1996.

Vol. 1647: D. Dias, P. Le Barz, Configuration Spaces over Hilbert Schemes and Applications. VII. 143 pages. 1996.

Vol. 1648: R. Dobrushin, P. Groeneboom, M. Ledoux, Lectures on Probability Theory and Statistics. Editor: P. Bernard. VIII, 300 pages. 1996.

Vol. 1649: S. Kumar, G. Laumon, U. Stuhler, Vector Bundles on Curves – New Directions. Cetraro, 1995. Editor: M. S. Narasimhan. VII, 193 pages. 1997.

Vol. 1650: J. Wildeshaus, Realizations of Polylogarithms. XI, 343 pages. 1997.

Vol. 1651: M. Drmota, R. F. Tichy, Sequences, Discrepancies and Applications. XIII, 503 pages. 1997.

Vol. 1652: S. Todorcevic, Topics in Topology. VIII, 153 pages. 1997.

Vol. 1653: R. Benedetti, C. Petronio, Branched Standard Spines of 3-manifolds. VIII, 132 pages. 1997.

Vol. 1654: R. W. Ghrist, P. J. Holmes, M. C. Sullivan, Knots and Links in Three-Dimensional Flows. X, 208 pages. 1997.

Vol. 1655: J. Azéma, M. Emery, M. Yor (Eds.), Séminaire de Probabilités XXXI. VIII, 329 pages. 1997.

Vol. 1656: B. Biais, T. Björk, J. Cvitanic, N. El Karoui, E. Jouini, J. C. Rochet, Financial Mathematics. Bressanone, 1996. Editor: W. J. Runggaldier. VII, 316 pages. 1997.

Vol. 1657: H. Reimann, The semi-simple zeta function of quaternionic Shimura varieties. IX, 143 pages. 1997.

Vol. 1658: A. Pumarino, J. A. Rodríguez, Coexistence and Persistence of Strange Attractors. VIII, 195 pages. 1997.

Vol. 1659: V, Kozlov, V. Maz'ya, Theory of a Higher-Order Sturm-Liouville Equation. XI, 140 pages. 1997.

Vol. 1660: M. Bardi, M. G. Crandall, L. C. Evans, H. M. Soner, P. E. Souganidis, Viscosity Solutions and Applications. Montecatini Terme, 1995. Editors: I. Capuzzo Dolcetta, P. L. Lions. IX, 259 pages. 1997.

Vol. 1661: A. Tralle, J. Oprea, Symplectic Manifolds with no Kähler Structure. VIII, 207 pages. 1997.

Vol. 1662: J. W. Rutter, Spaces of Homotopy Self-Equivalences – A Survey. IX, 170 pages. 1997.

Vol. 1663: Y. E. Karpeshina; Perturbation Theory for the Schrödinger Operator with a Periodic Potential. VII, 352 pages. 1997.

Vol. 1664: M. Väth, Ideal Spaces. V, 146 pages. 1997.

Vol. 1665: E. Giné, G. R. Grimmett, L. Saloff-Coste, Lectures on Probability Theory and Statistics 1996. Editor: P. Bernard. X, 424 pages, 1997.

Vol. 1666: M. van der Put, M. F. Singer, Galois Theory of Difference Equations. VII, 179 pages. 1997.

Vol. 1667: J. M. F. Castillo, M. González, Three-space Problems in Banach Space Theory. XII, 267 pages. 1997.

Vol. 1668: D. B. Dix, Large-Time Behavior of Solutions of Linear Dispersive Equations. XIV, 203 pages. 1997.

Vol. 1669: U. Kaiser, Link Theory in Manifolds. XIV, 167 pages. 1997.

Vol. 1670: J. W. Neuberger, Sobolev Gradients and Differential Equations. VIII, 150 pages. 1997.

Vol. 1671: S. Bouc, Green Functors and G-sets. VII, 342 pages. 1997.

Vol. 1672: S. Mandal, Projective Modules and Complete Intersections. VIII, 114 pages. 1997.

Vol. 1673: F. D. Grosshans, Algebraic Homogeneous Spaces and Invariant Theory. VI, 148 pages. 1997.

Vol. 1674: G. Klaas, C. R. Leedham-Green, W. Plesken, Linear Pro-p-Groups of Finite Width. VIII, 115 pages. 1997.

Vol. 1675: J. E. Yukich, Probability Theory of Classical Euclidean Optimization Problems. X, 152 pages. 1998.

Vol. 1676: P. Cembranos, J. Mendoza, Banach Spaces of Vector-Valued Functions. VIII, 118 pages. 1997.

Vol. 1677: N. Proskurin, Cubic Metaplectic Forms and Theta Functions. VIII, 196 pages. 1998.

Vol. 1678: O. Krupková, The Geometry of Ordinary Variational Equations. X, 251 pages. 1997.

Vol. 1679: K.-G. Grosse-Erdmann, The Blocking Technique. Weighted Mean Operators and Hardy's Inequality. IX, 114 pages. 1998.

Vol. 1680: K.-Z. Li, F. Oort, Moduli of Supersingular Abelian Varieties. V, 116 pages. 1998.

Vol. 1681: G. J. Wirsching, The Dynamical System Generated by the 3n+1 Function. VII, 158 pages. 1998.

Vol. 1682: H.-D. Alber, Materials with Memory. X, 166 pages. 1998.

Vol. 1683: A. Pomp, The Boundary-Domain Integral Method for Elliptic Systems. XVI, 163 pages. 1998.

Vol. 1684: C. A. Berenstein, P. F. Ebenfelt, S. G. Gindikin, S. Helgason, A. E. Tumanov, Integral Geometry, Radon Transforms and Complex Analysis. Firenze, 1996. Editors: E. Casadio Tarabusi, M. A. Picardello, G. Zampieri. VII, 160 pages. 1998.

Vol. 1685: S. König, A. Zimmermann, Derived Equivalences for Group Rings. X, 146 pages. 1998.

Vol. 1686: J. Azéma, M. Émery, M. Ledoux, M. Yor (Eds.), Séminaire de Probabilités XXXII. VI, 440 pages. 1998.

Vol. 1687: F. Bornemann, Homogenization in Time of Singularly Perturbed Mechanical Systems. XII, 156 pages. 1998.

Vol. 1688: S. Assing, W. Schmidt, Continuous Strong Markov Processes in Dimension One. XII, 137 page. 1998.

Vol. 1689: W. Fulton, P. Pragacz, Schubert Varieties and Degeneracy Loci. XI, 148 pages. 1998.

Vol. 1690: M. T. Barlow, D. Nualart, Lectures on Probability Theory and Statistics. Editor: P. Bernard. VIII, 237 pages. 1998.

Vol. 1691: R. Bezrukavnikov, M. Finkelberg, V. Schechtman, Factorizable Sheaves and Quantum Groups. X, 282 pages. 1998.

Vol. 1692: T. M. W. Eyre, Quantum Stochastic Calculus and Representations of Lie Superalgebras. IX, 138 pages. 1998.

Vol. 1743: L. Habermann, Riemannian Metrics of Constant Mass and Moduli Spaces of Conformal Structures. XII, 116 pages. 2000.

Vol. 1744: M. Kunze, Non-Smooth Dynamical Systems. X, 228 pages. 2000.

Vol. 1745: V. D. Milman, G. Schechtman, Geometric Aspects of Functional Analysis. VIII, 289 pages. 2000.

Vol. 1746: A. Degtyarev, I. Itenberg, V. Kharlamov, Real Enriques Surfaces. XVI, 259 pages. 2000.

Vol. 1747: L. W. Christensen, Gorenstein Dimensions. VIII, 204 pages. 2000.

Vol. 1748: M. Ruzicka, Electrorheological Fluids: Modeling and Mathematical Theory. XV, 176 pages. 2001.

Vol. 1749: M. Fuchs, G. Seregin, Variational Methods for Problems from Plasticity Theory and for Generalized Newtonian Fluids. VI, 269 pages. 2001.

Vol. 1750: B. Conrad, Grothendieck Duality and Base Change. X, 296 pages. 2001.

Vol. 1751: N. J. Cutland, Loeb Measures in Practice: Recent Advances. XI, 111 pages. 2001.

Vol. 1752: Y. V. Nesterenko, P. Philippon, Introduction to Algebraic Independence Theory. XIII, 256 pages. 2001.

Vol. 1753: A. I. Bobenko, U. Eitner, Painlevé Equations in the Differential Geometry of Surfaces. VI, 120 pages. 2001.

Vol. 1754: W. Bertram, The Geometry of Jordan and Lie Structures. XVI, 269 pages. 2001.

Vol. 1755: J. Azéma, M. Émery, M. Ledoux, M. Yor, Séminaire de Probabilités XXXV. VI, 427 pages. 2001.

Vol. 1756: P. E. Zhidkov, Korteweg de Vries and Nonlinear Schrödinger Equations: Qualitative Theory. VII, 147 pages. 2001.

Vol. 1757: R. R. Phelps, Lectures on Choquet's Theorem. VII, 124 pages. 2001.

Vol. 1758: N. Monod, Continuous Bounded Cohomology of Locally Compact Groups. X, 214 pages. 2001.

Vol. 1759: Y. Abe, K. Kopfermann, Toroidal Groups. VIII, 133 pages. 2001.

Vol. 1760: D. Filipović, Consistency Problems for Heath-Jarrow-Morton Interest Rate Models. VIII, 134 pages. 2001.

Vol. 1761: C. Adelmann, The Decomposition of Primes in Torsion Point Fields. VI, 142 pages. 2001.

Vol. 1762: S. Cerrai, Second Order PDE's in Finite and Infinite Dimension. IX, 330 pages. 2001.

Vol. 1763: J.-L. Loday, A. Frabetti, F. Chapoton, F. Goichot, Dialgebras and Related Operads. IV, 132 pages. 2001.

Vol. 1764: A. Cannas da Silva, Lectures on Symplectic Geometry. XII, 217 pages. 2001.

Vol. 1765: T. Kerler, V. V. Lyubashenko, Non-Semisimple Topological Quantum Field Theories for 3-Manifolds with Corners. VI, 379 pages. 2001.

Vol. 1766: H. Hennion, L. Hervé, Limit Theorems for Markov Chains and Stochastic Properties of Dynamical Systems by Quasi-Compactness. VIII, 145 pages. 2001.

Vol. 1767: J. Xiao, Holomorphic Q Classes. VIII, 112 pages. 2001.

Vol. 1768: M.J. Pflaum, Analytic and Geometric Study of Stratified Spaces. VIII, 230 pages. 2001.

Vol. 1769: M. Alberich-Carramiñana, Geometry of the Plane Cremona Maps. XVI, 257 pages. 2002.

Vol. 1770: H. Gluesing-Luerssen, Linear Delay-Differential Systems with Commensurate Delays: An Algebraic Approach. VIII, 176 pages. 2002.

Vol. 1771: M. Émery, M. Yor, Séminaire de Probabilités 1967-1980. A Selection in Martingale Theory. IX, 553 pages. 2002.

Vol. 1772: F. Burstall, D. Ferus, K. Leschke, F. Pedit, U. Pinkall, Conformal Geometry of Surfaces in S^4. VII, 89 pages. 2002.

Vol. 1773: Z. Arad, M. Muzychuk, Standard Integral Table Algebras Generated by a Non-real Element of Small Degree. X, 126 pages. 2002.

Vol. 1774: V. Runde, Lectures on Amenability. XIV, 296 pages. 2002.

Vol. 1775: W. H. Meeks, A. Ros, H. Rosenberg, The Global Theory of Minimal Surfaces in Flat Spaces. Martina Franca 1999. Editor: G. P. Pirola. X, 117 pages. 2002.

Vol. 1776: K. Behrend, C. Gomez, V. Tarasov, G. Tian, Quantum Comohology. Cetraro 1997. Editors: P. de Bartolomeis, B. Dubrovin, C. Reina. VIII, 319 pages. 2002.

Vol. 1777: E. García-Río, D. N. Kupeli, R. Vázquez-Lorenzo, Osserman Manifolds in Semi-Riemannian Geometry. XII, 166 pages. 2002.

Vol. 1778: H. Kiechle, Theory of K-Loops. X, 186 pages. 2002.

Vol. 1779: I. Chueshov, Monotone Random Systems. VIII, 234 pages. 2002.

Vol. 1780: J. H. Bruinier, Borcherds Products on O(2,1) and Chern Classes of Heegner Divisors. VIII, 152 pages. 2002.

Vol. 1781: E. Bolthausen, E. Perkins, A. van der Vaart, Lectures on Probability Theory and Statistics. Ecole d' Eté de Probabilités de Saint-Flour XXIX-1999. Editor: P. Bernard. VII, 480 pages. 2002.

Vol. 1782: C.-H. Chu, A. T.-M. Lau, Harmonic Functions on Groups and Fourier Algebras. VII, 100 pages. 2002.

Vol. 1783: L. Grüne, Asymptotic Behavior of Dynamical and Control Systems under Perturbation and Discretization. IX, 231 pages. 2002.

Vol. 1785: J. Arias de Reyna, Pointwise Convergence of Fourier Series. XVIII, 175 pages. 2002.

Recent Reprints and New Editions

Vol. 1200: V. D. Milman, G. Schechtman, Asymptotic Theory of Finite Dimensional Normed Spaces. 1986. – Corrected Second Printing. X, 156 pages. 2001.

Vol. 1618: G. Pisier, Similarity Problems and Completely Bounded Maps. 1995 – Second, Expanded Edition VII, 198 pages. 2001.

Vol. 1629: J. D. Moore, Lectures on Seiberg-Witten Invariants. 1997 – Second Edition. VIII, 121 pages. 2001.

Vol. 1638: P. Vanhaecke, Integrable Systems in the realm of Algebraic Geometry. 1996 – Second Edition. X, 256 pages. 2001.

Vol. 1702: J. Ma, J. Yong, Forward-Backward Stochastic Differential Equations and Their Applications. 1999. – Corrected Second Printing. XIII, 270 pages. 2000.